Proceedings in Life Sciences

Mechanisms of Protein Synthesis

Structure-Function Relations, Control Mechanisms,
and Evolutionary Aspects

Proceedings of the Symposium
on Molecular Mechanisms in Protein Synthesis
Held at Beyaz Köşk, Emirgan,
Bosphorus, Istanbul

Edited by E. Bermek

With 124 Figures

Springer-Verlag
Berlin Heidelberg New York Tokyo 1985

Prof. Dr. ENGIN BERMEK
Biyofizik Bilim Dali
Istanbul Tip Fakültesi
Istanbul Üniversitesi
Çapa, Istanbul, Türkiye

Cover illustration: The front cover illustrates a photograph of the 70S ribosome model, see p. 95

ISBN 3-540-13653-3 Springer-Verlag Berlin Heidelberg New York Tokyo
ISBN 0-387-13653-3 Springer-Verlag New York Heidelberg Berlin Tokyo

Library of Congress Cataloging in Publication Data. Main entry under title: Mechanisms of protein synthesis. (Proceedings in life sciences). Includes index. 1. Protein biosynthesis–Congresses. I. Bermek, E. (Engin), 1939-. II. Series. QP551.M368 1984 574.19'297 84-13979

Typesetting and printing: Beltz, Offsetdruck, Hemsbach/Bergstr.
Bookbinding: Brühlsche Universitätsdruckerei, Giessen
2131-3130-543210

Preface

This volume contains the papers presented at the international symposium on "Molecular Mechanisms in Protein Synthesis" held on September 26-27, 1983 at the Beyaz Köşk in Emirgan, Bosphorus, Istanbul.

The symposium aimed to create a medium for information exchange and discussions regarding the current developments in the area of protein synthesis. To ensure an informal yet scientifically stimulating and productive atmosphere providing opportunity for relaxed and speculative discussions, the number of presentations was limited to twenty and that of attendants to about sixty. The emphasis in the symposium was laid on structure-function relations in the prokaryotic protein synthesizing systems and on the control mechanisms of eukaryotic protein synthesis, in particular, during chain initiation. Other issues like evolutionary aspects of protein synthesis, translational components genes and proofreading were covered as well.

The manuscripts represent the extended accounts of the oral presentations, and it has been aimed with the concluding remarks at the end of the volume to give a summarizing view of the presentations and the discussions.

The symposium was sponsored by the Istanbul Faculty of Medicine and the Medical and Basic Sciences Research Groups of the Scientific and Technical Research Council of Turkey (TÜBITAK). The support of the Turkish Airlines, Koç Holding A.S., Oesterreichisches Generalkonsulat-Kulturinstitut, Rotary Club, SESA Elektronik San. Tic. A.S., Hoffmann-La Roche Company, Incekara A.Ş., Sandoz-Turkey, Turkish Glassworks Inc., Ünitay Export-Import, Kerman A.S., and Eczacibaşi Holding A.S., in Istanbul is gratefully acknowledged. I would also like to thank all those who have made with their help and contributions the organization of the symposium possible and particularly M. Metinsoy for her ever patient management of the secretarial tasks.

I wish that by propagating the reported findings and the views arisen during the symposium, this volume publication of which has been possible with the willing and competent collaboration of Springer-Verlag, Heidelberg, may provide new stimulus into different directions of research in protein synthesis.

Çapa, October 1984   E. Bermek

Contents

Proof-Reading in Translation

Regulation of Eukaryotic Chain Initiation

Eukaryotic Chain Elongation

Evolution and Selective Action of Translational Inhibitors

Contributors

You will find the adress at the beginning of the respective contribution

Barta, A. 40
Bermek, E. 170, 180
Bilgin-Aktar, N. 180
Cantor, C.R. 2
De Haro, C. 160
De Herreros, A.G. 160
Deng, H.Y. 23
DiSegni, G. 120
Fasano, D. 50
Faulhammer, H. 61
Gökhan, N. 170
Guesnet, J. 50
Hardesty, B. 23
Hopfield, J.J. 106
Hui, C.F. 2
Kaempfer, R. 120
Kan, B. 170
Kang, C. 2
Kanigür, G. 170
Küchler, E. 40
Lai, C.Y. 106
Levin D.H. 144
Lill, R. 76
London, I.M. 144
Matts, R.L. 144
Murphy, R.F. 2

Nurten, R. 180
Ochoa, S. 160
Odabaş, Ü.G. 180
Odom, O.W. 23
Parlato, G. 50
Parmeggiani, A. 50
Paulsen, H. 76
Petryshyn, R. 145
Picone, D. 50
Pizzano, R. 50
Robbins, D. 23
Robertson, J.M. 76
Rosen, H. 120
Rychlik, W. 23
Sayhan, O.Z. 180
Serdyuk, I.N. 92
Spirin, A.S. 92
Sprinzl, M. 61
Steiner, G. 40
Şimşek, M. 194
Thomas, N.S. 144
Tiryaki, D. 170
Vazquez, D. 201
Voorma, H.O. 135
Wintermeyer, W. 76
Wollenzien, P.L. 2
Yamane, T. 106

Ribosome Structure

RNA Structure, Free and on the Ribosome, as Revealed by Chemical and Enzymatic Studies

P.L. WOLLENZIEN, C.F. HUI, C. KANG, R.F. MURPHY, and C.R. CANTOR[1]

1. Introduction

At present, various chemical and enzymatic probes are the experimental techniques capable of providing large amounts of data on the structure and arrangement of large RNA molecules. Direct reactivity measurements yield information on accessibility, while cross-linking provides information on proximity. Taken together with secondary structure estimates derived from analysis of base pairing potential and phylogenetic variations, the results of chemical and enzymatic studies can help evaluate specific models for RNA structures and interactions. It is unlikely that a precise three-dimensional structure could result from this approach. However, it should be possible, through accumulation of a large enough data set, to provide reasonably accurate secondary structures and approximate models of how the secondary structure is arranged in space, alone, or in concert with proteins. A unique advantage of this approach is that it is suited to the detection of conformational changes in RNA molecules. Thus, it will be very useful in establishing functional aspects of altered RNA structures where they exist.

Escherichia coli ribosomal RNAs are the ideal system in which to develop and evaluate particular probes for large RNA structure. They can readily be isolated either as free rRNAs or as ribosomal subunits; the primary structures are completely known (Brosius et al. 1978; Carbon et al. 1979; Brosius et al. 1980). In addition, there are considerable data suggesting that, near physiological conditions, the free RNAs can fold into compact, three-dimensional structures similar to their structures on the ribosome (Vasiliev et al. 1978; Spirin et al. 1979; Allen and Wong 1978; Vasiliev and Zalite, 1980). Detailed secondary structure models have been presented (Noller and Woese 1981; Stiegler et al. 1981a; Zwieb et al. 1981; Noller et al. 1981; Glotz et al. 1981; Branlant et al. 1981) and while these are still evolving it is clear that at the present time we know more about the secondary structures of the 16S rRNA and 23S rRNA than any other RNA of comparable sizes.

The rRNAs have been identified as the structural core for both ribosomal subunits (Stuhrmann et al. 1977; Stuhrmann et al. 1978; Spirin et al. 1979). Since there is considerable experimental evidence for structural changes within the ribosome during protein translation (see Zamir et al. 1974) and for altered topographies of the rRNA under different conditions (Chapman and Noller 1977; Hogan and Noller 1978; Herr and Noller 1979; Brow and Noller 1983), it is reasonable to suspect that changes in the state of the

---

[1] Department of Human Genetics and Development, College of Physicians and Surgeons, Columbia University, New York, New York 10032, USA

Mechanisms of Protein Synthesis
ed. by Bermek
© Springer-Verlag Berlin Heidelberg 1984

rRNA are intimately connected to the process of translation. Noller and Woese (1981) have argued, based on evolutionary considerations, that the primordial ribosome may have been RNA alone. In this view the ribosomal proteins are relegated to the secondary role of fine tuning the speed and fidelity of protein translation. In addition, there are several sites in the rRNAs directly implicated in ribosome function. (see Thompson and Hearst 1983b, Küchler et al. 1984). It thus remains of considerable interest to characterize the structure of ribosomal RNAs as fully as possible. It is also of great interest to understand more about the structure of tRNAs and mRNAs when these are bound on the ribosome in various functional states.

The least that direct reactivity measurements could reveal about the structure of an RNA, whether free or in a ribonucleoprotein complex, is the identity of residues on the surface. For a molecule the size of 16S rRNA this could be an appreciable fraction of the whole. The most simpleminded calculations reveal that roughly half of the 1542 residues of the 16S rRNA should be on the surface of the free RNA it if is quite compact; more if it is at all extended.

Most direct chemical and enzymatic studies can also provide information about the secondary structure of the residues involved; however, this pattern is superimposed on the pattern of accessible sites. Examples are enzymes specific for single or double strands and chemical reactions with atoms involved in base pairing in double helices. Unfortunately, we have no way at present of evaluating the effect of RNA tertiary structures on such reaction specificity. Thus, all such inferences about secondary structures and surface accessibility have to be considered with some caution.

Since failure to react with an enzyme or chemical probe can always be the result of an idiosyncratic local environment or conformation, only positive results, i.e., reactivity, are generally interpretable.

To get the maximum value from accessibility studies one needs many probes, preferably with different sizes to try to infer the location of any narrow crevices or channels. It is critical that large amounts of data must be generated easily and even more clear that this data must be as error free as possible. Even a few erroneous positives, i.e., reactive groups that are actually located internally, can badly distort any models built from the data. This is because there is no simple way to cross-check the data. The amount of information needed to reconstruct even crude three-dimensional models is enormous and it is always tempting to try to use all the data available to refine the models. This is fine if all the data is accurate, but it can be disasterous otherwise.

Cross-linking is a direct way of determining the proximity between two residues so long as one can guarantee that the formation of the cross-link has not artifactually brought together two residues normally far apart. Even in this case one knows that the residues can be near each other, but not necessarily in the native state. The advantage of photoreagents for cross-linking is the short time scale of their reactivity. This virtually insures that only residues in normal contact will be cross-linked. However, the current availability of RNA photo cross-linkers is pretty much limited to psoralens which are specific for base paired residues. To probe other types of proximity one must resort to chemical reagents. One of the most useful nucleic acid cross-linkers is the thiol-containing glyoxal derivative GbzCyn$_2$Ac (Expert-Bezancon and Hayes 1980). This reacts with single stranded guanines analogously with glyoxal. Reduction generates free thiols which can then be cross-linked either by direct oxidation or by bridging with bis thiol reagents, such as bis maleimides. In this way, one can cross-link nearby single stranded guanine residues, indicating features of the tertiary structure.

In principle, a sufficient battery of cross-linking results could provide a complete three-dimensional structure accurate to about the size of the reagents themselves, typically 10 Å. This would be enormously valuable and might well approach what could be available from X-ray diffraction on such large and potentially floppy structures. There is no way to obtain the correct mirror handedness by cross-linking alone, but this difficulty would be resolved by correlation of the possible structures with protein-RNA cross-linking and positions of proteins determined by immune electron microscopy. However, to achieve the goal of a complete three-dimensional structure at high resolution would require hundreds of cross-links. It would be relatively easy to obtain cross-links for residue pairs accessible to externally added reagents. Indeed, most currently available information on the 16S rRNA has involved the use of rather bulky reagents and, thus, the results obtained are surely a view of the structure heavily biased towards the more accessible regions. For internal proximity measurements one has to resort to more difficult "zero length" cross-links, such as those provided by direct irradiation through a variety of direct and indirect photochemical mechanisms.

Again, as in the case of accessibility measurements, maximally effective cross-linking will involve the collection of large amounts of data, as error free as possible. The availability of extensive secondary structure data will greatly reduce the number of cross-links needed to define the three-dimensional structure and will also provide some useful redundancy checking.

The great potential of cross-linking is offset by the amount of effort that is potentially involved. The approach for RNA is still in its infancy. The 16S rRNA is the most intensively studied species and, thus far, less than 50 distinct cross-links within the 16S rRNA are indicated by currently used reagents (Zwieb and Brimacombe 1980; Wollenzien and Cantor 1982a; Thompson and Hearst 1983a; Expert-Bezancon et al. 1983; Wollenzien et al. 1984). Few of these cross-links have yet been mapped at the level of individual residues. To achieve the potential impact of cross-linking we will have to have many more useful reagents and rapid and accurate ways of analyzing their reaction patterns. In this paper we focus on summarizing the current state of the art in studying psoralen cross-links in 16S rRNA. Many of the obstacles to using psoralens rapidly and effectively are now being overcome. The solutions should offer considerable guidance to those who will use psoralens or other cross-linkers in studies of any large RNA.

2. Electron Microscopic Techniques for Studying Psoralen Cross-linking

Many of the products produced by cross-linkers will always be pairs of residues that lie reasonably close in the nucleotide sequence (Zwieb and Brimacombe 1980; Turner et al. 1982; Expert-Bezancon et al. 1983; Thompson and Hearst 1983a). While this sort of data is useful confirmation of secondary structure models in many cases, it is not often particularly novel or informative. Instead, what one hopes in a cross-linking study, is to see products involving pairs of residues distant in the primary structure. These can provide either tests of long-distance secondary structure contacts, vital in discriminating between different models, or information about RNA tertiary structure contacts. In either case the long-distance contacts provide information that is invaluable for constructing three-dimensional structure models, and it is information that is almost impossible to obtain by other existing methods.

The major advantage of analysis of cross-link locations by electron microscopy is that one is forced to focus on cross-links between distant residues. The basic idea of the method is very simple and is shown schematially in Fig. 1. A population of RNAs, hopefully homogeneous, in solution or on the ribosome, is treated with a cross-linking agent. The excess agent is removed or inactivated, any proteins present are removed, and then the RNA is subjected to a condition which completely and irreversibly removes all secondary structure. We use treatment with formamide followed by heating in the presence of formaldehyde (Wollenzien and Cantor 1982a). The resulting molecules are then spread on a grid with a cytochrome c technique, and then stained with uranyl acetate and shadowed with platinum and palladium in order to visualize them in the electron microscope.

Molecules containing no cross-links will almost always appear as worm-like rods; loops or circles caused by accidental crossover of two parts of the same molecule are observed less than 3 % of the time for RNAs the size of 16S rRNA (Wollenzien et al. 1984). A typical field of uncross-linked 16S rRNA molecules is shown in Fig. 2 (top). In contrast, with cross-linked molecules, the presence of the extra covalent bond imposed by each cross-link insures that a loop will remain after the RNA structure is denatured. Thus, cross-links between residues distant along the chain will be immediately apparent in the electron microscope by the appearance of looped structures. Figure 2 (bottom) shows a typical field generated by psoralen cross-linking of free 16S rRNA. It is apparent that under the conditions used, roughly a quarter of the RNA molecules show loops. A wide variety of different loop and tail configurations is evident.

Large fields of molecules can be scanned visually to obtain statistics on the frequency of occurrence of cross-links which produce unique and easily discriminated loop patterns. Examples are full circles which indicate cross-links between residues near both ends of the RNA, and "lollipop structures", molecules with a small loop at one end,

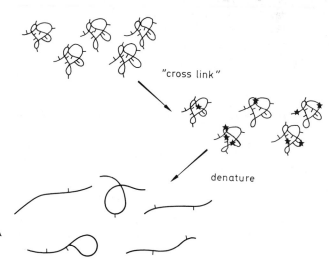

Fig. 1. The analysis of RNA cross-linking by electron microscopy

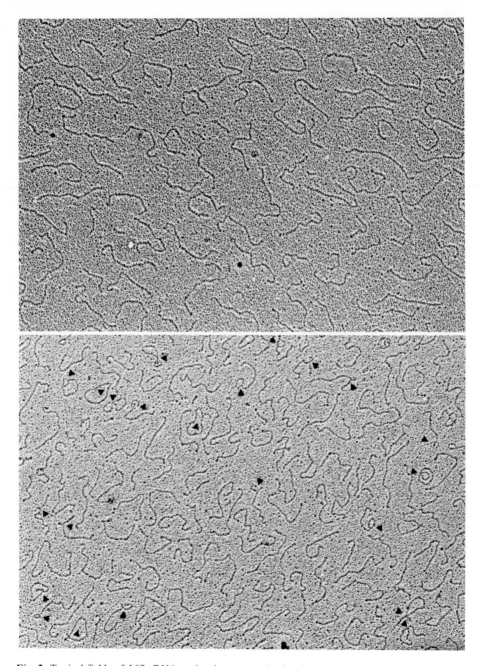

Fig. 2. Typical fields of 16S rRNA molecules as seen in the electron microscope by the spreading techniques in current use. At the *top* is a field of uncross-linked RNAs; at the *bottom* a field of RNAs cross-linked by irradiation in solution with the psoralen AMT. The *arrowheads* show molecules with easily detected cross-linked loops

indicating a cross-link between an end and an internal region somewhere between the middle and that end. For more precise identification of specific cross-links, a much more tedious procedure is required. Individual molecules containing any looped structure are photographed. The lengths of their loops and tails are measured by analyzing the image with a digital planimeter. Because different molecules are stretched to different extents during the preparation of the sample for microscopy, there is significant variation in the quantitative appearance even of molecules with identical cross-links. This is the major limitation in the resolution of specific cross-linked species and the determination of the location of the cross-links. The problem can be partially overcome by observing many examples of each type of cross-linked molecule. Each molecule makes up one member of a large data set that is then analyzed statistically, as will be described later.

A compromise must be made between the yield of molecules with interesting cross-linked features and the ability to analyze these molecules accurately. If too few molecules contain cross-links, the real data will be obscured by noise from the occasional accidental molecular crossover during spreading for microscopy. If too many cross-links are present in a molecule, the pattern of loops that results cannot be traced uniquely and, therefore, all information is lost.

With psoralens, and probably most other cross-linkers, there is another serious problem. Since the desired cross-links between residues distant in the primary structure are not necessarily favored reaction products, by the time these products have appeared at significant yield, the molecules containing them may have one or more additional cross-links between residues nearby in the sequence. If these not visible directly in the microscope, they would alter the apparent length of the RNA chain by a length equal to the distance between the cross-links. Since the distortion can occur, in principle, in any tail or loop, the result would lead to serious deterioration of the sharpness of the lengths characterizing each particular cross-linked loop. A check on this problem can be made by measuring the apparent shortening of contour length of cross-linked molecules compared to control uncross-linked molecules (Wollenzien and Cantor 1982a; Wollenzien et al. 1984).

To overcome the distortion caused by "hidden" cross-links, it is necessary to work at rather low levels of cross-linking. This means that very large numbers of molecules must be examined and one must be scrupulously careful to perform spreading under conditions that minimize accidental crossovers. Techniques are available for fractionating cross-linked species away from uncross-linked molecules and these can be used where necessary to enrich the population of desired products prior to electron microscopy (Wollenzien and Cantor 1982a). There are several consequences of performing cross-linking at low levels that must be kept in mind. As shown schematically in Fig. 1, the curious result of a cross-linking procedure is to take a sample that was once homogeneous and convert it to a heterogeneous mixture of products. Almost any method used for subsequent analysis has to contend with this heterogeneity. This is why microscopy, as a single molecule technique, has an overwhelming potential advantage over any other technique. Representation in the heterogeneous mixture will be determined mainly by the reactivity of the particularly cross-linked residues under specified conditions.

The occurrence of open, untangled structures is a constant feature for all the loops that have been seen by psoralen cross-linking of 16S rRNA and other RNAs. It is fortunate since it greatly facilitates the quantitative analysis of the electron microscopic images. It also immediately provides an important piece of structural information: there are no structures with knotted topologies formed by residues within any of the loops (Cantor et al. 1980). RNA helices form one complete turn every 11 base pairs and many possible structures involving multiple turns of helix formed by interrupted stretches of sequence would lead to topologically knotted chain configurations. Once the region was closed into a covalent circle, these knots would be topologically trapped and the resulting molecules would appear to have complex extra looped structures within the circles. The fact that these are never observed forms a powerful structural constraint that will be a critical test of potential three-dimensional folding patterns.

Any inspection of the cross-linked molecules in Fig. 2 will immediately reveal that there is no way to distinguish which end of the molecule is which. The location of each loop has an uncertain polarity. One must have additional information to determine which end of the molecule is the 3' end and which is the 5' end. Earlier psoralen cross-linking studies (Wollenzien et al. 1979a) had to rely on rather indirect ways of doing this (Wollenzien et al. 1979b) and the results led to (what we know now) the incorrect orientation of some of the cross-links made in the free 16S rRNA. What was needed was a way of directly marking one end of the RNA in the electron microscope with a technique that would survive the rather rigorous denaturation conditions required for proper spreading. This demanded a covalent label and the technique that was finally developed took advantage of the unique photochemical properties of psoralens (Fig. 3).

When aminomethyl trioxsalen (AMT) or hydroxymethyl trioxsalen (HMT) are irradiated at 390 nm, near their long wavelength absorption edge, instead of forming cross-links in double stranded DNA or RNA, they form a monoadduct with just one of the strands (Chatterjee and Cantor 1978). At any time later, a second shorter wavelength irradiation, typically at 360 nm, will convert the monoadduct into a cross-link providing that the sequence at that point contains the alternating purine-pyrimidine dinucleotide required for the second photoreaction. This allows the end labeling procedure described schematically in Fig. 4 (Wollenzien and Cantor 1982b).

Appropriate recombinant DNA techniques are used to obtain a plasmid containing part of the 16S rRNA gene in a known position. Digestion of this plasmid DNA with restriction endonucleases and subsequent size fractionation by polyacrylamide gel electrophoresis is used to prepare a restriction fragment containing several hundred base pairs of sequence coding for regions near one end of the 16S rRNA. This fragment is treated

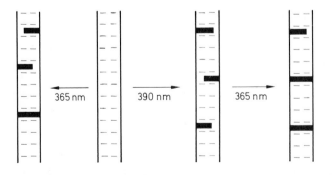

Fig. 3. Psoralen photo-chemistry. Shown schematically are the production of monoadducts by long wavelength irradiation and their conversion to cross-links by shorter wavelengths. Direct short wavelength irradiation produces a mixture of cross-links and mono-adducts

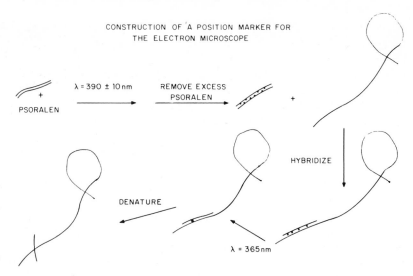

Fig. 4. The technique used to label one end of a cross-linked 16S rRNA in the electron microscope. See the text for further details

with psoralen to produce an average of several monoadducts, but no cross-links. The fragment is denatured by heating and then hybridized to 16S rRNA containing cross-links previously placed there by any agent desired. When the RNA-DNA hybrid is reirradiated at 360 nm, the psoralens form an average of one cross-link, covalently attaching the DNA to the RNA. Now when the sample is denatured and examined in the microscope, the extra DNA fragment appears as a short line that clearly marks the location of the end to which it is attached. The identity of that end is known from the nucleotide sequence of the fragment. Typical molecules labeled in this way are shown in Fig. 5. The technique is simple and unambigous.

In previous studies, the absolute orientation of cross-linked RNA molecules in the electron microscope image was not known. The best that one could do to classify molecules was to measure the position of cross-links from the shortest tail. Using this data each molecule can be represented as a point in a triangular two-dimensional region as shown in Fig. 6a (Wollenzien and Cantor 1982a). Once data on many molecules is accumulated, distinct cross-linked species will manifest themselves as a cluster of many molecules within a small area of the triangular region.

In our recent studies the absolute orientation of the cross-linked molecules is known from the labeling method described above. Thus, each molecule containing a single cross-linked loop can be described by two parameters, an estimate of the distance from the beginning, X, and end, Y, of the loop to the 5, end. From this definition X is always less than Y. Each molecule is then represented as a point in a triangular two-dimensional region as shown in Fig. 6b. Note that this region is twice as large as the corresponding region when unoriented molecules are used. This increase in area results in a markedly enhanced ability to discriminate between many different cross-linked species. Note that the data on unoriented molecules can be generated from data on oriented molecules by folding the triangular region of the latter in half. Thus, once individual cross-linked features are identified with an oriented sample it will be possible

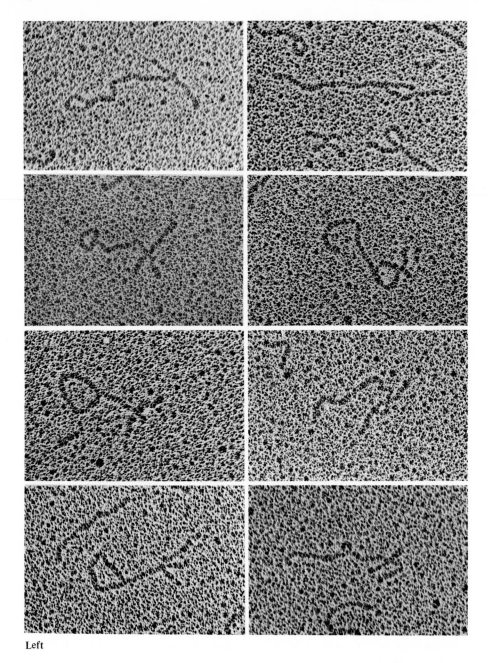

Left

Fig. 5. Examples of end labeled molecules. At *left* are images of the molecules; at *right* are tracings that interpret each of these images: RNA is shown as a *solid line* and DNA as a *dashed line*. The bar indicates 0.5 microns

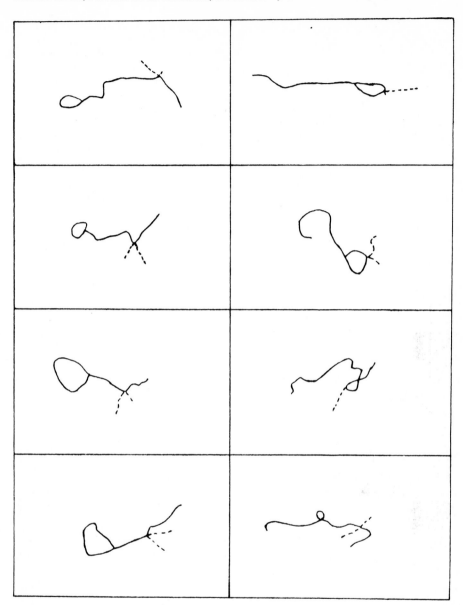

Fig. 5. Right

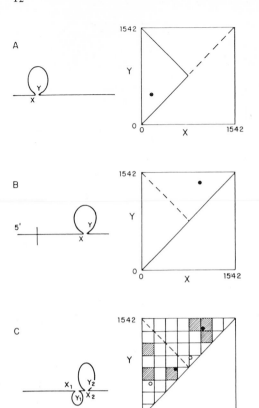

Fig. 6 A-C. Representation of a cross-linked molecule by a triangular plot. **A** If the polarity of the molecule is not known, X, the position of the start of the loop, is measured from the closer terminus and Y, the position of the end of the loop, is also measured from the same terminus; **B** if the polarity is known, X and Y are both measured from the 5' terminus; **C** once the distribution of the most frequent cross-links is established on the triangular plot, this information can be used to determine the polarity of molecules which have two cross-linked loops; the correct orientation *(solid circles)* fit the pattern much better than the incorrect orientation *(open circles)*

in some cases to uniquely assign molecules in an unoriented sample. This can greatly increase the number of molecules available for analysis since it is far less trouble to prepare samples without end labels.

We have stressed thus far the desirability of working with molecules containing as few cross-links as possible and most of the molecules used for our studies contain only a single visible cross-linked loop in the electron microscope. However, in preparing such samples, one will always produce an appreciable fraction of molecules with two visible cross-linked loops since at low extents of reaction one is simply randomly sampling multiple potential targets. For molecules containing two nonoverlapping loops and an end label, one can assign the values $X < Y < X' < Y'$ uniquely to the two cross-links. Thus, such molecules are useful data. Even if the molecule is not oriented, in some cases it can be used. If one of the cross-linked loops corresponds to a structure that has already been assigned uniquely in an oriented sample, it provides the orientation of the molecule and thus allows the location of the second cross-linked loop to be assigned uniquely (Wollenzien et al. 1984). This is the same kind of indirect approach that was used in earlier studies (Wollenzien et al. 1979a), but now it is being applied in a far more restrictive fashion.

The final stage in electron microscopic analysis of cross-linking is to take plots, such as shown schematically in Fig. 6, and subject them to statistical analysis to determine the number of independent cross-linking events needed to fit the observed distribution of molecules to within the accuracy justifiable by statistical criteria. Such fits require estimates of the errors in the determination of each apparent image in the microscope and also require assumptions about the distributions of such errors. A detailed discussion of the methods used is beyond the scope of this article. However, it is useful to point out that, from our practical experience, a minimum of 25-50 molecules per component is needed before reasonable convincing fits can be obtained. Examples of some of the results obtained with psoralen cross-linked 16S rRNA samples (Wollenzien and Cantor 1982a; Wollenzien et al. 1984) are shown in Fig. 7. These will be discussed more fully in the next section.

While our discussion has focused on psoralen cross-links, most of the comments are applicable to any RNA cross-linking scheme. Thus, $GbzCyn_2Ac$ cross-links and direct UV cross-links can be analyzed in exactly the same manner as outlined above. At best the electron microscope allows the cross-link to be localized to about ± 25 bases. In most cases such accuracy allows one to determine the organization of domains in the secondary or tertiary structure of the RNA. For tests of specific secondary structure models or for construction of precise tertiary structure models it is necessary to have more accurate cross-link locations.

Psoralens prefer to react with U's rather than C's, and G-U pairs at the end of duplexes or unpaired U's appear to be especially reactive regions. Two pyrimidines are required to make the cross-link and, thus, two U's are the most likely target. It might be possible to make educated guesses about the actual sites of cross-linking if the reactivity of different base paired arrangements were known. However, at the present time it does not seem worthwhile to interpret the electron microscopic results to this extent.

The situation with the $GbzCyn_2Ac$ is somewhat different. Single stranded guanines are the expected targets and there is every reason to believe that the reactivity of $GbzCyn_2Ac$ with individual guanines should mirror that of kethoxal or glyoxal under similar conditions. Since the pattern of kethoxal and glyoxal reactivity has been mapped at the sequence level, the identity of residues likely to be prime targets for $GbzCyn_2Ac$ is known. This allows the preliminary identification of the regions involved in $GbzCyn_2Ac$ cross-linking (Expert-Bezancon and Hayes 1980; Expert-Bezancon and Wollenzien 1984).

3. Results of Psoralen Cross-linking of 16S rRNA

It is not our intent here to give a detailed summary of all of the psoralen cross-links that have been mapped in the 16S rRNA under various conditions. The available data is continually being upgraded as more and more molecules are included, and details are available elsewhere (Wollenzien and Cantor 1982a; Chu et al. 1983; Wollenzien et al. 1984). Here we would like to compare the overall pattern of the psoralen cross-linking data seen in free 16S rRNA and in the inactivated conformation of the 30S ribosomal subunit. We study the inactivated conformation because the accessibility of the activated

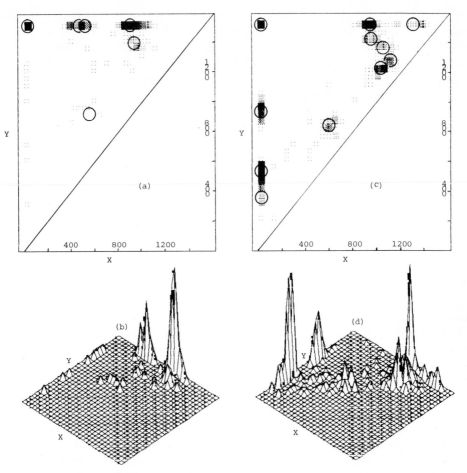

Fig. 7 a-d. Triangular plots of oriented cross-linked 16S rRNA molecules. **a** 16S rRNA psoralen cross-linked in inactive 30S subunits, density plot; **b** same data as in **a** on a perspective plot; **c** psoralen cross-linked as free 16S rRNA, density plot; **d** same data as in **c** on a perspective plot. The plots contain a total of 113 and 523 molecules, respectively. In **a** the density levels are graded from 1 *(lightest)* to more than 8 *(darkest)* and in **b** the density levels are graded from 2 *(lightest)* to more than 16 *(darkest)*. The perspective plots have been rotated so that the data at the left axis can be seen

conformation of the 30S particle is so low that it is difficult to accumulate a sufficient population of cross-linked molecules for electron microscopy (Thammana et al. 1979). We believe that studies on the inactivated 30S particle are relevant for understanding RNA structure in functioning ribosomes because the inactive 30S may resemble the 30S moiety of the 70S particle (Zamir et al. 1974; Ginsburg and Zamir 1975).

Figure 7 summarizes the psoralen cross-links seen in oriented molecules of free 16S rRNA and inactive 30S subunits. Several features of the figure should be apparent after close inspection. Some of the peaks in the two triangular maps are similar. In fact the pattern of cross-links of the 16S rRNA in the inactive 30S subunit forms a subset of that produced in free 16S rRNA (Thammana et al. 1979; Wollenzien et al. 1984). Thus, the identity of some cross-linked species is presumably the same. However, the relative heights of many of the peaks are quite different. In the inactive 30S particle almost all of the prominent peaks indicate cross-links in which one of the residues is near the 3' end of the RNA (Thammana et al. 1979; Wollenzien and Cantor 1982a). This is consistent with a variety of chemical and enzymatic studies of the accessibility of RNA in the inactive 30S subunit (see Noller and Woese 1981; Vassilenko 1981; Teterina et al. 1980). These studies indicate that the 3' terminal region of the 16S rRNA is quite exposed and reactive, while most of the remainder of the RNA is fairly inaccessible. The results illustrate the concern raised earlier that at low extents of cross-linking the data will be highly biased towards the most reactive or accessible residues. If one did not know ahead of time about the differential accessibility of the RNA in the particle one could easily be misled to the conclusion that the structures of the free and ribosome bound RNA were quite different.

The pattern of reaction of the free 16S rRNA shows a number of prominent cross-links that involve residues near the 5' end of the RNA and in other interior regions of the 16S rRNA in addition to those cross-links prominent in both free and ribosome-bound RNA involving the 3' end. The pattern of cross-linking in the free 16S rRNA provides strong confirmation for the general secondary structure models that have evolved from other studies on the 16S rRNA (Wollenzien et al. 1984). Table 1 shows major long-distance contacts which appear in a number of different secondary structure models and agreement with some of the psoralen cross-links. Considering the large number of long-distance contacts that are hypothetically possible, the agreement is quite gratifying. However, there are additional strong psoralen cross-links that do not correspond to postulated secondary structure features. These presumably indicate the occurrence of additional base paired contacts. According to current custom in RNA structural studies we call such contacts tertiary structure. They represent base paired regions too short to be established by sequence or phylogenetic analysis or base "pairs" that are not conventional A-U and G-C. Examples are the wide variety of base-base interactions that supplement normal base pairs in the tertiary structure of tRNAs.

A striking aspect of the data in Fig. 7 is the lack of any significant population of cross-links in the upper left hand sector of the triangular map. This corresponds to contacts between residues 0-500 and 1000-1500. Thus, except for the very 3' end of the RNA there appears to be little interaction between the first and lasts thirds of the molecule. This immediately suggests that the RNA is divided into domains not just at the secondary structure level, as indicated by various models, but also at the tertiary structure level. Results shown elsewhere indicate that $GbzCyn_2 Ac$ cross-linking also fails to produce many products in this region of the map, supporting the conclusion from the psoralen results (Wollenzien et al. 1984; Expert-Bezancon and Wollenzien 1984).

The functional significance of domains in the RNA structure is mostly a matter for speculation at present. In proteins, domains have been shown to move with respect to each other. This is likely to be true for RNA also and, thus, the possibility of significant RNA motions in one or more stages of protein synthesis must be seriously considered. The most tantalizing hints of function from the pattern of psoralen cross-linking involve the 3' end of the 16S rRNA, from residues 1400 to 1542. Parts of this region seem

Table 1. Comparison of secondary structure models with psoralen cross-linked 16S rRNA[a]

Complementarity	Cross-link (in free RNA)	Cross-link (RNA in subunit)
27- 37 / 547- 556	EPs 5'x 540 (7.9 %)	Not detected
17- 20 / 915- 918	EPs 5'x 930 (6.6 %)	Not detected
39- 47 / 394- 403 52- 58 / 354- 359	EPs 5'x 360 (4.8 %)	Not detected
946-955 / 1225-1235 984-990 / 1214-1221	EPs 1000x1230 (1.3 %)	Not detected
564-570 / 880- 886 576-580 / 761- 765 584-587 / 754- 757	EPs 580x 840 (0.9 %)	EPs 550x 870 (0.1 %)
926-933 / 1384-1391	EPs 930x1430 (0.9 %)	EPs 950x1400 (0.4 %)
937-943 / 1340-1345	EPs 950x1340 (0.8 %)	Not detected
Not predicted	EPs 920x 3' (7.2 %)	EPs 930x 3' (21 %)
Not predicted	EPs 5'x 3' (1.8 %)[b]	EPs 5'x 3' (1.5 %)[b]
Not predicted	EPs 1300x 3' (1.6 %)	Not detected
Not predicted	- - -[c]	EPs 510x 3' (2.2 %) EPs 450x 3' (1.4 %)
Not predicted	EPs 1080x1280 (0.5 %)	Not detected

[a] Results are summarized from Wollenzien et al. (1984) for psoralen cross-linked free 16S rRNA and from Wollenzien and Cantor (1982a) for 16S rRNA psoralen cross-linked in inactivated 30S subunits. The preface EPs indicates the electron microscopic identification of psoralen cross-links. Cross-link frequencies reported are percentages of total number of molecules

[b] Molecules containing EPs 5'x3' appear as circularized 16S RNA without any small tails; since they appear symmetrical they have not been included in the hybridized-oriented data sets. The position of this cross-link is indicated in Fig. 7a and c

[c] Molecules containing cross-link EPs 420x3' have been identified in histograms of unoriented molecules at a frequency of 1.6 %. The tentative orientation results from the similarity to cross-links EPs 450x3' and EPs 510x3' made in the inactivated 30S subunit

extraordinarily well conserved through various prokaryotes and eukaryotes (Noller and Woese 1981; Stiegler et al. 1981b) and it is known to be intimately involved in protein synthesis activity by a variety of studies. The message recognition region is here (Shine and Dalgarno 1975; Steitz and Jakes 1975); the modified bases involved in kasugamycin sensitivity are here (Helser et al. 1973); the anticodon of P-site bound tRNA is in proximity to this region (Taylor et al. 1981; Ofengand and Lion 1981). A number of specific roles for other conserved residues in this region have recently been postulated (Thompson and Hearst 1983b).

The striking features of the psoralen cross-linking results is the diversity of contacts made by the 3' end of the RNA in both the free molecule and in the inactive 30S particle. This suggests that the 3' end may be capable of existing in a number of very different discrete structures. The psoralen cross-linking will trap these structures individually.

Their existence would imply the possibility of major structural rearrangements of the 3' end that could clearly play an important role in critical steps in protein synthesis. It should eventually be possible to test this possibility by analyzing the pattern of cross-links in molecules with more than one loop visible in the electron microscope. If there is a set of different structures, certain pairs of cross-links should tend to appear in the same molecule while other pairs will never coexist. If the cross-linking is sampling a single homogeneous structure, then the pattern of multiple cross-links will just reflect the joint probabilities of their individual independent appearance.

4. Analysis of Cross-links at the Sequence Level

A major limitation in all of the studies described above is the imprecision of cross-link assignment in the electron microscope. Knowledge of the exact sequence location of each psoralen or $GbzCyn_2Ac$ cross-link would be invaluable in the test of secondary structure models and the construction of three-dimensional models.

Direct chemical or enzymatic analysis of the locations of cross-links is a commonly used technique for protein-protein contacts. After fragmentation and separation of fragments in a typical fingerprinting analysis, the cross-link is broken and the identity of the two short cross-linked fragments revealed by diagonal separation techniques. Recently this kind of an approach has also been developed for UV cross-linked 16S RNA (Zwieb and Brimacombe 1980), for psoralen cross-linked RNA (Thompson and Hearst 1983a), and for $GbzCyn_2Ac$ cross-linked 16S RNA (Expert-Bezancon et al. 1983). The power of this technique is that, in principle, it provides the highest possible resolution information, namely, the actual cross-linked residues. The difficulty is that, in addition to being tedious, there is no easy way, in existing methods, to know the frequency of the cross-links that are sequenced or to select for potentially interesting cross-links between residues distant in the primary structure.

We have sought to devise a strategy that could be used to identify residues near or at the site of cross-linking without the need to fragment the RNA as in a typical finger-print analysis. We also desired a method that could selectively examine cross-links between residues distant in the primary structure since, as we have argued above, these are the most interesting products. While the details of our method for cross-link sequencing are still evolving, enough progress has now been made to outline the general scheme. It is an approach very different from those described previously.

RNAs containing cross-links between residues distant in the primary structure can be isolated free from most other cross-linked species by gel electrophoresis in the presence of a strong denaturant like formamide (Wollenzien and Cantor 1982a). Such separations afford good resolution of RNAs with different cross-links. In general, molecules with large loops migrate with lower mobility, although the position of the loop in the molecule is also important. Thus, it is possible to obtain a sample highly enriched in one, or at most a few psoralen cross-links. This is absolutely essential, since in the procedure described below the two residues involved in each cross-link are evaluated separately and one must not lose the correlation between them. The purified cross-linked RNA sample is then hybridized to an end labeled complementary DNA fragment as shown schematically in Fig. 8. The fragment is chosen so that it overlaps the general region of the cross-link, as determined by prior electron microscopy. The DNA fragment is short enough so that it can be completely sequenced by a single polyacrylamide gel electrophoresis. The RNA-DNA hybrid is subjected to mung bean nuclease digestion. This nuclease is single strand specific and it should nick the DNA at or near the bases complementary to the cross-linked residues, since, as shown in Fig. 8 these residues are unavailable for base pairing to DNA in the hybrid by virtue of the cross-link within the RNA strand.

An experimental test of this method is shown in Fig. 9. The RNA contains the major psoralen cross-link seen in inactivated 30S particles. The DNA is the XmaIII-XmaIII restriction fragment of *rrnB* corresponding to residues to 899-1141 of the 16S rRNA. When a control hybrid between uncross-linked RNA and the DNA is subjected to mung bean nuclease digestion a series of bands are seen. These presumably reflect occasional nicking of the DNA at regions where there is some tendency for the RNA-DNA duplex to melt or breath under the conditions of the digestion. Note, however, that when cross-linked RNA is used, in addition to this background of bands, a number of distinct new products are observed. The position of these is in excellent agreement with that expected from microscopy. While more work needs to be done to confirm this result it is an encouraging hint that we may soon have a method that could rapidly assign the sequence location of many 16S rRNA cross-links.

5. The Arrangement of Ribosome-bound Message

Some RNA sequences cannot be easily cross-linked or probed with informative chemical reagents. We have long been interested in the structure of ribosome-bound mRNA and tRNA. Here one must use very gentle methods to avoid displacing the noncovalently bound RNAs from their binding sites. Footprinting the pattern of accessible residues has been successfully reported for tRNA by both tritium exchange and chemical methods. Much less is known about mRNA. Indeed there is substantial disagreement among several early studies as to the length of the message bound to the ribosomes (Gupta et al. 1970; Pieczenik et al. 1974; Steitz 1979). There is even substantial disagreement on the length of poly (U) bound to the 70S ribosome (Takanami and Zubay 1964; Castles and Singer 1969; Pochon et al. 1977).

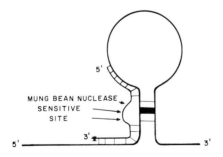

Fig. 8. A technique for finding the location of cross-links in an RNA by generating nuclease nicks in a hybrid with a complementary DNA fragment

Fig. 9. A typical set of cross-link sequencing data for the sample described in the text. At *left* is the mung bean nuclease digestion pattern of DNA hybridized to uncross-linked RNA; at *right* the pattern when the RNA contains the major (nominally 930-1540) cross-link seen in the inactivated 30 S particle. The new bands seen in the cross-link contain 27 to 30 nucleotides

While there are obvious reasons to study the structure of natural messages on the ribosome we chose to start with the synthetic message poly (U). Since this material is nearly devoid of secondary structure its pattern of reactivity on the ribosome would have to reflect ribosome interactions free from any complications arising from the structure of the RNA itself. Since there are no good chemical probes for uracil, we were led to examine various enzymatic probes. Ultimately the one that was particularly useful is ribonuclease T2. This is a relatively nonspecific enzyme.

The results of a typical experiment are shown in Fig. 10. Here uniformly ^{32}P-labeled poly (U) was bound to 70S tight couple ribosomes in the presence of excess yeast tRNA[phe]. After T2 nuclease digestion to a point where the resulting product mixture was relatively constant with time, chain lengths of the ribosome-protected poly (U) fragments were analyzed by polyacrylamide gel electrophoresis. If the poly (U) were uniformly protected by ribosome binding one would expect to see a very sharp distribution of product lengths. The maximum length would demark the maximum length of mRNA at which some contact with the ribosome were felt while the length distribution would indicate the penetrability of the edges of the mRNA binding site to the nuclease.

The actual pattern seen is much broader and much more complicated than expected. The longest fragments seen in any experiment are about 52 nucleotides. This estimate for the binding site size is in really excellent agreement with a recent estimate by decay of fluorescence anisotropy (Pochon et al. 1977). The presence of several maxima in the length distribution implies that there are several sites along the ribosome-bound message that are still accessible to the nuclease. Further studies should reveal the actual position of these sites. However, the fact that they exist at all implies that the ribosome-bound message must reach the surface of the 70S particle at several places. Whether these places represent loops of mRNA actually free from the ribosome remains to be seen. However, given the ease and rapidity of modern ladder sequencing techniques it should be possible to probe much more fully the conformation of RNA substrates in the ribosome in a variety of functional states. It is fortunate that RNAs appear to play many of the key roles in protein synthesis since with currently available techniques they are far easier to study than the ribosomal proteins.

Acknowledgement. This work was supported by a grant from the USPHS. The collaboration of Alain Bezancon and Y.G. Chu in work that was just touched upon in this overview, is gratefully appreciated.

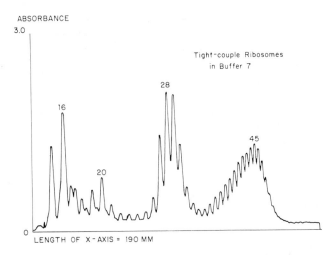

Fig. 10. A typical length distribution of poly (U) fragments after T2 ribonuclease digestion of 70S tight couples containing excess tRNA[Phe] . The *arrows* indicate sizes determined by band counting from the position of known oligo (U) standards

References

Allen SH, Wong K-P (1978) The hydrodynamic and spectroscopic properties of 16S RNA from *Escherichia coli* ribosome in reconstitution buffer. J Biol Chem 255: 8759-8766

Branlant C, Krol A, Machatt MA, Pouyet J, Ebel JP, Edwards K, Kossel H (1981) Primary and secondary structures of *Escherichia coli* MRE 600 23S ribosomal RNA comparison with models of secondary structure for maize chloroplast 23S rRNA and for large portions of mouse and human 16S mitochondrial rRNAs. Nucl Acids Res 9: 4303-4324

Brosius J, Palmer ML, Kennedy PJ, Noller HF (1978) Complete nucleotide sequence of 16S ribosomal RNA gene from *Escherichia coli*. Proc Natl Acad Sci USA 75: 4801-4805

Brosius J, Dull TJ, Noller HF (1980) Complete nucleotide sequence of a 23S ribosomal RNA gene from *Escherichia coli*. Proc Natl Acad Sci USA 77: 201-204

Brow DA, Noller HF (1983) Protection of ribosomal RNA from Kethoxal in polyribosomes. Implication of specific sites in ribosomal function. J Mol Biol 163: 27-46

Carbon P, Ehresmann C, Ehresmann B, Ebel J-P (1979) The complete nucleotide sequence of the ribosomal 16S RNA from *Escherichia coli*. Experimental details and cistron heterogeneities. Eur J Biochem 100: 399-410

Castles JJ, Singer MF (1969) Degradation of polyuridylic acid by ribonuclease II: Protection by ribosomes. J Mol Biol 40: 1-17

Chapman NM, Noller HF (1977) Protection of specific sites in 16S from chemical modification by association of 30S and 50S ribosomes J Mol Biol 109: 131-149

Chu YG, Wollenzien PL, Cantor CR (to be published 1983) J Biomolec Struct and Dynam

Expert-Bezancon A, Hayes D (1980) Synthesis and properties of N-acetyl-N'-(P-glyoxylbenzoyl) cystamine, a new reagent for RNA-RNA and RNA-protein cross-linking. Eur J Biochem 103: 365-375

Expert-Bezancon A, Milet M, Carbon P (1983) Eur J Biochem 136: 267-274

Expert-Bezancon A, Wollenzien PL (1984) manuscript in preparation

Ginsburg I, Zamir A (1975) Characterization of different conformational forms of 30S ribosomal subunits in isolated and associated states: possible correlations between structure and function. J Mol Biol 93: 465-476

Glotz C, Zweib C, Brimacombe R, Edwards K, Kossel H (1981) Secondary structure of the large subunit ribosomes RNA from *Escherichia coli,* Zea mays chloroplast and human and mouse mitochondrial ribosomal. Nucl Acids Res 9: 3287-3306

Gupta SL, Chen J, Schaefer L, Lengyel P, Weissman SM (1970) Nucleotide sequence of a ribosome attachment site of bacteriophage f2 RNA. Biochem Biophys Res Commun 39: 883-888

Helser TL, Davies JE, Dahlberg JF (1973) Mechanism of kasugamycin resistance in *Escherichia coli*. Nature New Biol 235: 6-9

Herr W, Noller HF (1979) Protection of specific sites in 23S and 5S RNA from chemical modification by association of 30S and 50S ribosomes. J Mol Biol 130: 421-432

Hogan JJ, Noller HF (1978) Altered topography of 16S RNA in the inactive form of *Escherichia coli* 30S ribosomal subunits. Biochemistry 17: 587-593

Küchler E, Steiner G, Barta A (1984) this volume

Noller HF, Woese CR (1981) Secondary structure of 16S ribosomal RNA. Science (Wash DC) 212: 403-411

Noller HF, Kop J, Wheaton V et al. (1981) Secondary structure model for 23S ribosomal RNA. Nucl Acids Res 9: 6167-6189

Ofengand J, Liou R (1981) Correct codon-anticodon base-pairing at the 5'-anticodon position blocks covalent cross-linking between transfer ribonucleic acid and 16S RNA at the ribosomal P site. Biochemistry 20: 552-559

Pieczenik G, Model P, Robertson HD (1974) Sequence and symmetry in ribosome binding sites of bacteriophage fl RNA. J Mol Biol 90: 191-214

Pochon F, Armand B, Lavalette D (1977) Rotational diffusion of Escherichia coli ribosomes II. Ribosome attachment to polyuridylic acid. Biochemie (Paris) 59: 785-788

Shine J, Dalgarno L (1975) Determinant of cistron specificity in bacterial ribosomes. Nature (Lond) 254: 34-38

Spirin AS, Serdyuk IN, Serpungin JL, Vasiliev VD (1979) Quaternary structure of the ribosomal 30S subunit-model and its experimental testing. Proc Natl Acad Sci USA 76: 4867-4871

Steitz JA (1979) In: Goldberger RF (ed) Biological Regulation and Development, vol 1. Plenum, New York, pp 349-399

Steitz JA, Jakes K (1975) How ribosomes select initiator regions in mRNA: base pair formation between the 3' terminus of 16S rRNA and the mRNA during initiation of protein synthesis in *Escherichia coli*. Proc Natl Acad Sci USA 72: 4734-4738

Stiegler P, Carbon P, Zucker M, Ebel JP, Ehresmann C (1981a) Structural organization of the 16S ribosomal RNA from *E. coli*: Topography and secondary structure. Nucl Acids Res 9: 2153-2172

Stiegler P, Carbon P, Ebel JP, Ehresmann C (1981b) A general secondary-structure model for procaryotic and eucaryotic RNAs of the small ribosomal subunits. Eur J Biochem 120: 487-495

Stuhrmann HB, Koch MHJ, Parfait R, Haas J, Crichton RR (1977) Shape of 50S subunit of *Escherichia coli* ribosomes. Proc Natl Acad Sci USA 74: 2316-2320

Stuhrmann HB, Koch MHJ, Parfait R, Haas J, Ibel K, Crichton RR (1978) Determination of the distribution of protein and nucleic acid in the 70S ribosomes of *Escherichia coli* and their 30S subunits by neutron scattering. J Mol Biol 119: 203-212

Takanami M, Zubay G (1964) An estimate of the size of the ribosomal site for messenger-RNA binding. Proc Natl Acad Sci USA 51: 834-839

Taylor BH, Prince JB, Ofengand J and Zimmermann RA (1981) Nonanucleotide sequence from 16S ribonucleic acid at the peptidyl transfer ribonucleic acid binding site of the *Escherichia coli* ribosome. Biochemistry 20: 7581-7588

Teterina NL, Kopylov AM, Bogdanov AA (1980) Segments of the 16S RNA located on the surface of the 30S subunits of *Escherichia coli* ribosomes. FEBS Lett 116: 265-268

Thammana P, Cantor CR, Wollenzien PL, Hearst JE (1979) Crosslinking studies on the organization of the 16S ribosomal RNA within the 30S *Escherichia coli* ribosomal subunit. J Mol Biol 135: 271-283

Thompson JF, Hearst JE (1983a) Structure of *E. coli* 16S RNA elucidated by psoralen crosslinking. Cell 32: 1355-1365

Thompson JF, Hearst JE (1983b) Structure function relations in *E. coli* 16S RNA. Cell 39: 19-24

Turner S, Thompson JF, Hearst JE, Noller HF (1982) Identification of a site of psoralen crosslinking in *E. coli* 16S ribosomal RNA. Nucl Acids Res 10: 2839-2849

Vasiliev VD, Selivanova OM, Koteliansky VE (1978) Specific selfpacking of the ribosomal 16S RNA. FEBS Lett 95: 273-276

Vasiliev VD, Zalite OM (1980) Specific compact selfpacking of the ribosomal 23S RNA. FEBS Lett 121: 101-104

Vassilenko SK, Carbon P, Ebel JP, Ehresmann C (1981) Topography of 16S RNA in 30S subunits and 70S ribosomes accessibility to cobra venom ribonuclease. J Mol Biol 152: 699-721

Wollenzien PL, Hearst JE, Thammana P, Cantor CR (1979a) Basepairing between distant regions of the *Escherichia coli* 16S ribosomal RNA in solution. J Mol Biol 135: 255-269

Wollenzien PL, Hearst JE, Squires C, Squires C (1979b) Determining the polarity of the map of crosslinked interactions in *Escherichia coli* 16S ribosomal RNA. J Mol Biol 135: 285-292

Wollenzien PL, Cantor CR (1982a) Gel electrophoretic technique for separating crosslinked RNAs. Application to improved electron microscopic analysis of psoralen crosslinked 16S ribosomal RNA. J Mol Biol 159: 151-166

Wollenzien PL, Cantor CR (1982b) Marking the polarity of RNA molecules for electron microscopy by covalent attachment of psoralen-DNA restriction fragments. Proc Natl Acad Sci USA 79: 3940-3944

Wollenzien PL, Expert-Bezancon A, Murphy RF, Cantor CR (1984) manuscript in preparation

Zamir A, Miskin R, Vogel Z, Elson D (1974) The inactivation and reactivation of *Escherichia coli* ribosomes. Methods Enzymol 30F: 406-426

Zwieb C, Brimacombe R (1980) Localization of a series of intra-RNA cross-links in 16S RNA, induced by ultraviolet irradiation of *Escherichia coli* 30S ribosomal subunits. Nucl Acids Res 8: 2397-2411

Zwieb C, Glotz C, Brimacombe R (1981) Secondary structure comparisons between small subunit ribosomal RNA molecules from six different species. Nucl Acids Res 9: 3621-3640

Studies of *E. coli* Ribosomes Involving Fluorescence Techniques

B. HARDESTY, O.W. ODOM, W. RYCHLIK, D. ROBBINS, and H.Y. DENG[1]

1. Introduction

Most models of the *Escherichia coli* ribosome and its subunits [see Figs. 2, 4, and 5 in Wittmann (1983)] are based primarily on images obtained by electron microscopy. Information about the spatial arrangements of the individual ribosomal proteins and rRNA[2] has come mainly from immune electron microscopy, neutron scattering, and protein-protein cross-linking experiments. Techniques involving fluorescence provide another powerful approach for determining ribosome structure and function. One of the most useful procedures involves determination of distances between pairs of probes by nonradiative singlet-singlet energy transfer. Though the useful range for energy transfer with commonly used donor and acceptor probes is relatively narrow (about 20 Å to 80 Å), this range is compatible with the size of most ribosomal proteins and tRNA. An advantage of this method is that a single fluorophore can be covalently attached to a specific known position, for instance a cysteine in a protein or a terminal nucleotide in an RNA molecule.

2. The Methods Employed

Experiments in which nonradiative energy transfer is used determine the distance between fluorophores relative to R_0, the distance at which energy transfer proceeds at half of its maximum rate. Evaluation of R_0 for a given pair of fluorophores has been discussed in previous publications (Odom et al. 1980; Robbins et al. 1981). Experimentally, the efficiency of energy transfer, E, is determined by measuring either the difference in fluorescence intensity or in fluorescence lifetime of the donor in the presence and absence of the acceptor. The fluorescence lifetime method potentially is independent of fluorophore concentration, whereas fluorescence intensity is a concentration-dependent parameter. Determination of energy transfer by comparing fluorescence intensity of the donor in samples containing only the donor and the donor-acceptor pair, respectively, requires assumptions or knowledge of the relative concentration of the donor in the two samples and of the proportion of donor probes that are actually paired with the acceptor in the sample containing both probes. It is desirable to have

---

[1] Clayton Foundation Biochemical Institute, Department of Chemistry, University of Texas, Austin, Texas 78712, USA
[2] See List of Abbreviations

Mechanisms of Protein Synthesis
ed. by Bermek
© Springer-Verlag Berlin Heidelberg 1984

the proportion of donors which are paired with acceptors as high as possible, preferably approaching unity so that a correction for the unpaired donor fluorophore in the doubly-labeled sample is not needed. However, in some cases it is not possible to obtain 100 % labeling with fluorophores. Methods have been devised to overcome the problem of unpaired donor fluorophore and differences in concentration between singly- and doubly-labeled samples. A useful approach to the problem of incomplete donor-acceptor pairing involves measuring fluorescence intensity before and after separation of probes by digestion of the sample with degradative enzymes (Epe et al. 1983).

3. Ribosomal RNAs and Proteins: Labeling and Biological Activity

Table 1 lists ribosomal components and other molecules involved in protein synthesis that have been labeled successfully with retention of biological activity in our laboratory. To label the 3' end of RNAs we have found that periodate oxidation of the cis diol groups of the terminal ribose followed by reaction of the resulting aldehyde with a thiosemicarbazide of the fluorophore of choice usually gives good results (Odom et al. 1980). The reaction sequence is shown in Fig. 1A. A similar reaction can be carried out with hydrazides. Generally these reactions provide a high degree of labeling, usually approaching 100 %, yield a covalent bond that is stable to most experimental manipulations, and are highly selective in the site of reaction. A potential difficulty with the procedure is that internal 3' ends caused by internal breaks in double-stranded regions of RNA molecules may also be labeled. This becomes a serious problem requiring special attention with large RNA molecules, such as 16S and particularly 23S rRNA.

Proteins may be labeled by reaction of their free amino groups or sulfhydryl groups. Amino group reactions generally result in labeling at a number of different sites

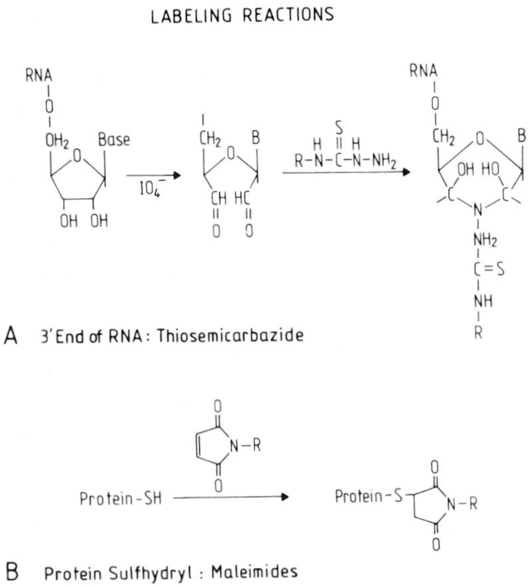

LABELING REACTIONS

A   3' End of RNA: Thiosemicarbazide

B   Protein Sulfhydryl : Maleimides

Fig. 1 A-B. Reactions for labeling RNA and proteins. **A** 3' end of RNA; **B** cysteine sulfhydryl groups of proteins

Table 1. Protein synthesis components which have been labeled with fluorescent probes with retention of biological activity[a]

Component	Position of label	Fluorophores used	Comments on biological activity[b]
rRNA			
16S	3' end	ETS, FTS	Active; less active with MS2-RNA
5S	3' end	ETS, FTS	Active
23S	3' end	ETS, FTS	Active
Proteins			
S1	-SH	FM, PM, CPM	Active; less active with MS2-RNA
S21	-SH	FM, CPM	Active in MS2-RNA-directed binding of fMet-tRNA$_f$
L11	-SH	FM, CPM	Active
L28	-SH	FM, CPM	Active
tRNA^{Phe}	3' end	NMA, ETS, Eos^H, FTS	Active in P-site binding
	X base	FITC, EITC, Fluram	Active in all tRNA reactions
	DHU bases	NMA, ETS, Eos^H, FTS	Poorly aminoacylated; active in P-site binding
	anticodon (Y base)	PF	
	X-link		
	(Srd$_8$-C$_{13}$)	X-link itself	Active in all tRNA reactions
IF-3	-SH	FM, CPM	Binds well to 30S subunits, activity may be reduced
EF-Tu (*Th. thermophilus*)	-SH	FM, CPM	Active in all EF-Tu reactions

[a] All components were isolated from *E. coli* except Tu

[b] Biological activity for rRNA and ribosomal proteins was assayed in poly (U)-directed polyphenylalanine synthesis

on the protein molecule; however, the sulfhydryl group provides a better opportunity for specific labeling. Many proteins, including a number of *E. coli* ribosomal proteins have only one or two cysteine residues, as shown in Table 2. We prefer reaction of a cysteine sulfhydryl group with a maleimide derivative of a fluorophore, as shown in Fig. 1B. At neutral pH maleimides generally react with strong preference for sulfhydryl groups (Liu 1977), although significant reaction with amino groups may occur under alkaline conditions. The sulfhydryl reaction with maleimides can be carried out under denaturing conditions, as in 7 *M* urea or 6 *M* guanidine·HCl, to allow labeling of sulf-hydryl groups that are inaccessible in nondenatured proteins.

We have been able to carry out efficient labeling of many of the sulfhydryl ribo-somal proteins, at least under denaturing conditions using the reactions indicated above. However, with the exception of S1, which can be removed efficiently from ribosomes by affinity chromatography on poly(U)-Sepharose (Subramanian et al. 1981), consid-erable difficulty was encountered initially in reconstituting the labeled proteins into ribosomal subunits. This difficulty may involve either of two types of problems. With some of the proteins, the relatively bulky fluorophore itself may interfere with binding of the protein or subsequent steps in reassembly of the subunits. This appears to be the situation with S4, S8, and S17 that bind early during 30S subunit assembly (Nomura and Held 1974), although use of labeled S4 has been reported (Epe et al. 1983). The other problem that has been partially overcome is to obtain subunits or mixtures of subunit proteins that lack the specific protein of interest. The so-called brute force approach requires subunit reconstitution from purified individual proteins or groups of proteins lacking the labeled protein. In some cases partial reconstitution can be carried out from LiCl cores of ribosomal subunits. Although this approach has been used with considerable success for studies involving neutron scattering (Moore 1980; Nierhaus et al. 1983), it requires technical resources that may be beyond the capacity of smaller research groups. A second approach that we have used successfully involves affinity chromatography to remove the protein of interest from subunits or mixtures of subunit proteins. S1 can be quantitatively removed from either mixtures of ribosomal proteins or intact 30S subunits by chromatography on a poly(U) column (Subramanian et al. 1981). S1 binds tightly to polynucleotides (K_b = $9\times10^6 M^{-1}$ for U_{12}; Linde et al. 1979) under conditions of low ionic strength, but can be eluted from a poly(U) column with 7 *M* urea, thus providing a convenient and efficient way to prepare both the protein and S1-deficient 30S ribosomal subunits.

Affinity chromatography also can be carried out with antibodies against the protein of interest. We have used this approach with antibodies provided by G. Stöffler (Berlin) and his co-workers to remove S2 or L11 from mixtures of 30S or 50S ribosomal pro-teins, respectively, with satisfactory results for the latter. However, only rather limited amounts of subunit protein can be processed even with a column prepared from rela-tively large amounts of antibody which must have high specificity for only the protein of interest.

A third and, where applicable, most useful approach involves the use of ribosomes from mutant strains of *E. coli* that lack or have a modified protein that can be easily removed from their ribosomes. E. Dabbs (Berlin) has a collection of such strains which include mutants for cysteine-containing proteins L11, L28 (Dabbs 1979), S21 and L27 (E. Dabbs, personal communication). Although some of the mutants grow very slowly

Table 2. -SH ribosomal proteins

Protein	# -SH	Cys position[a]	Total amino acids[a]	Moli wt. 10^{-3}	Axial ratio[b]	Protein location[c]	X-linking[d]
S1	2	292, 349	557	61.2	10	Platform region	S 3, 10, 18
2	1	86	240	26.6	5-7.5	Body opposite platform	S 3, 5, 8
4	1	31	203	23.1	Elongated	Platform	S 3, 5, 6, 8, 9, 12, 13, 17
8	1	126	129	14.0	2.3	Neck	S 11, 13, 15
11	2	69, 120	128	13.7		Lower head	S 8, 13, 21
12	4	26, 33, 52, 103	123	13.6		Center head	S 3, 4, 13, 20
13	1	63	117	13.0	2.0	Top head	S 4, 5, 7, 8, 11, 17, 19, 21
14	1	83	98	11.2		Platform side of head	S 19, 21
17	2	58, 63	83	9.6	3.5-4	Top head	S 4, 13
18	1	10	74	8.9	8-12	Lower body	S 1, 21
21	1	22	86	9.5	8-9.4	Platform	S 11, 13, 14
L2	2	5, 187	272	29.7		Base of c.p.[e]	L 5, 6, 7/12, 9, 10, 11, 17
5	1	86	178	20.2		Central body	L 2, 7/12, 11, 23, 25, 31
6	1	124	176	18.8	4.2	Central body	L 2, 7/12, 10, 11, 17
10	1	69	165	14.9	4-5	Central body	L 2, 6, 7/12, 11, 13, 14, 21, 31, 32
11	1	38	141	13.2		Central body	L 2, 4, 5, 6, 7/12, 10, 14, 17, 21
14	1	21	120	14.4		Central body	L 11, 17
17	1	100	127	9.0		Lower body	L 2, 5, 6, 11, 13, 14, 21, 27, 31, 32
27	1	52	84	8.9		Base of l.p.[e]	L 29
28	1	4	77				L 9
31	4	16, 18, 37, 40	62	7.0			L 1, 5, 10, 17, 18, 22, 23

[a] Sequence data provided by Dr. Wittmann-Liebold
[b] Wittmann et al. 1980
[c] Stöffler et al. 1980
[d] Traut et al. 1980
[e] c.p., central protuberance; l.p., lateral protuberance

and require considerable care to culture in large amounts, they can provide a relatively easy and convenient way to produce deficient ribosomes.

In addition to the ribosomal proteins, we have worked with elongation factor EF-Tu and initiation factor IF-3. Both contain cysteine residues that can be labeled by reaction with maleimides. IF-3 has a single sulfhydryl group (Brauer and Wittmann-Liebold 1977) that can be labeled without loss of activity. EF-Tu as well as EF-G from *E. coli* contain reactive cysteine residues, but both are inactivated by reaction with maleimides. However, EF-Tu from the extreme thermophile, *Thermus thermophilus,* contains two cysteine residues that can be labeled after reduction (Nakamura et al. 1978). The enzyme from the thermophile appears to be fully active with ribosomes and other components from *E. coli* (Arai et al. 1978a, b), can be easily isolated in relatively large amounts, and has proven to be remarkably stable and easy to work with experimentally.

Also given in Table 1 is the biological activity of the labeled macromolecules compared to unmodified controls. As determined by poly(U)-dependent polyphenylalanine synthesis, ribosomal RNAs that are labeled at their 3' ends with most fluorophores retain close to 100 % activity compared to unmodified RNA both reconstituted into ribosomal particles. Labeled tRNA^{Phe} with a fluorophore at any one of the five positions indicated retains its capacity to bind to ribosomes. However, a probe on the dihydrouridine loop appears to reduce or eliminate its capacity to be aminoacylated. Of course tRNA^{Phe} labeled at its 3' end cannot be aminoacylated.

4. Measured Distances

Distances between ribosomal components or tRNA that have been measured in our laboratory are summarized in Table 3. With the exception of the 5S RNA-16S RNA pair, the distances between probes on the 3' ends of the rRNAs are too large to be measured reliably, greater than about 65 Å using fluorescein to eosin as the energy donor-acceptor pair. However, about 44 % energy transfer corresponding to 51-59 Å was observed between probes on the 3' ends of 5S RNA and 16S RNA (Odom et al. 1980). These results are consistent with the positions of the 3' ends of rRNA as determined by immune electron microscopy (Stöffler-Meilicke et al. 1981).

Elongation factor EF-Tu is of special importance in the peptide elongation cycle in that it forms a ternary complex with aminoacyl-tRNA and GTP and thereby promotes binding of aminoacyl-tRNA into the ribosomal A site. GTP hydrolysis with subsequent release of EF-Tu·GDP normally occurs upon interaction of the ternary complex with a 70S ribosome. EF-Tu is not released from the ribosome if a nonhydrolyzable analogue is substituted for GTP.

Thermus thermophilus EF-Tu, labeled with maleimide derivatives after reduction, was used to study the location and mechanism of action of the factor on ribosomes (Rychlik et al. 1983). In the presence of a nonhydrolyzable analogue of GTP, a decrease in fluorescence from the coumarin derivative of the factor is observed as it is bound to ribosomes. This decrease in fluorescence was used to determine a binding constant of $(3 \pm 1.2) \times 10^6 \; M^{-1}$ for the reaction. Reconstituted 70S ribosomes containing either

Table 3. Distances between ribosomal components determined by non-radiative energy transfer

Components	Distance[b] (Å)	Reference
Proteins and rRNAs[a]		
5S RNA and 23S RNA	> 65	Odom et al. (1980)
5S RNA and 16S RNA	51-59	"
16S RNA and 23 S RNA	> 70	"
EF-Tu and S1	59-71	Rychlik et al. (1983)
EF-Tu and 16S	77-87	"
EF-Tu and 5S	67-74	"
EF-Tu and 23S	68-78	"
-SH groups of S1	24-31	Odom et al. (1984)
P site tRNA^{Phe} and 16S RNA[c]		
tRNA$^{Phe}_{XLR}$ [d]	53-60	Robbins et al. (1981)
3' end labeled	67-74	"
anticodon loop	> 61	"
A site tRNA^{Phe} and 16S RNA		
tRNA$^{Phe}_{XLR}$	> 60	Robbins & Hardesty (1983)
X base	77-88	"
anticodon loop	> 56	"

[a]The fluorescent probes were attached to the 3' ends of ribosomal RNAs and to the sulfhydryl groups of S1 and EF-Tu

[b]The range of distances given was determined from fluorescence polarization data by the method of Haas et al. (1978)

[c]Average values using several combinations of probes

[d]XLR is the Srd_8-C_{13} cross-link that has been reduced with $NaBH_4$

labeled S1 or an rRNA labeled at its 3' end were used for energy transfer experiments with the coumarin derivative of EF-Tu. The distance between the probe on the factor bound to the ribosome and the 3' ends of 16S RNA, 5S RNA, 23S RNA, or S1 were calculated to be 82, 70, 73, and 62-68 Å, respectively. These data appear to be consistent with an EF-Tu binding site near the base of the L7/L12 stalk on the 50S ribosomal subunit.

S1, the largest ribosomal protein, has two cysteine residues at positions 292 and 349. The thiol group of the latter, but not the former can be readily alkylated by maleimides in the native protein so that a monoalkylated derivative at cysteine 349 can be formed under nondenaturing conditions. The less reactive thiol of cysteine 292 can be alkylated in denaturing concentrations of urea or guanidine·HCl. These differences in thiol reactivity were used to label the cysteines with energy donor and acceptor probes with which the distance between the probes was estimated to be 24-31 Å or less (Odom et al. 1984). No difference in this distance could be detected if the protein was bound to 30S ribosomal subunits. Distance measurements involving other ribosomal proteins have been impaired by difficulty in preparing mixtures of ribosomal proteins or subunits with which the labeled protein of interest could be reconstituted. However, such studies are now underway using the procedures described above.

One of the most promising experimental approaches to understand ribosomal function involves distance measurements from different points on tRNA bound to ribosomes in different sites. Such studies were carried out with ribosomes containing 3' labeled 16S RNA and tRNA^{Phe} labeled at three different specific points (Robbins et al. 1981): E. coli and yeast tRNA^{Phe} were labeled at the 3' end; the Y base in position 37 at the 3' side of the anticodon of yeast tRNA^{Phe} was replaced by proflavin or 1-amino-anthracene; E. coli tRNA^{Phe} was photochemically cross-linked between 4-thiouracil at position 8 and cytosine at position 13 (Srd_8-C_{13} cross-link). The cross-linked structure formed is fluorescent after reduction. These three species provide probes near the top, middle, and bottom of the three-dimensional tRNA^{Phe} structure. The labeled tRNAs were bound to the P site of 70S ribosomes, and energy transfer from the fluorophore in the modified tRNA to the probe at the 3' end of 16S RNA was determined from fluorescence lifetimes. The distances to the 3' end of 16S RNA were: 3' end of tRNA, 67-74 Å; Srd_8-C_{13} cross-link, 53-60 Å; anticodon loop of tRNA, >61 Å.

These results were unanticipated since the evidence for the involvement of the 3'-terminal region of the 16S RNA in peptide initiation (Steitz 1980) and in peptide elongation (Ofengand et al. 1980) suggests that the 3' terminus of the 16S RNA might be near the decoding region of the small ribosomal subunit. However, the energy transfer data indicate that the 3' rRNA terminus is actually located closer to the center of the tRNA than to the Y base position in the anticodon loop.

In another set of experiments (Robbins and Hardesty 1983) the distance from three points on Phe-tRNA in the ribosomal acceptor (A) site to a probe on the 3' end of 16S RNA was measured. The tRNA probes used were on the anticodon loop of yeast tRNA^{Phe}, on the X-base of E. coli tRNA^{Phe}, and the Srd_8-C_{13} cross-link. The ribosomal P site was blocked with unlabeled, deacylated tRNA^{Phe} and then Phe-tRNA labeled at one of the points indicated was bound to the ribosome with EF-Tu and GTP. In contrast to the results described above for tRNA in the P site, the distance between the probes on the 16S RNA and the cross-link was too great to be measured, 60 Å or more (Table 3). The corresponding distance to the X-base probe was 77-88 Å. These data are consistent with the anticipated model in which tRNA in the A site is farther from the 3' end of 16S RNA than the tRNA in the P site.

5. The Effect of Aurodox on T. thermophilus EF-Tu

Parmeggiani and co-workers have conducted extensive studies on the mechanism by which kirromycin inhibits protein synthesis (review, Parmeggiani and Sander 1980). Kirromycin (mocimycin) and its N-methylated derivative, aurodox or antibiotic X5108 (Liu et al. 1977), bind to EF-Tu from E. coli and appear to have identical effects on protein synthesis in the E. coli cell-free system (Wolf et al. 1974). Kirromycin was reported to bind to EF-Tu in a 1:1 molar ratio causing an increase in the affinity of the factor for GTP by two orders of magnitude to a value approaching that of the EF-Tu·GDP complex (K_{eq} = 1 nM). A concomitant large increase in the exchange rate between free and bound GDP was observed (Blumenthal et al. 1977; Fasano et al. 1978). The conclusion was reached that binding of kirromycin to EF-Tu causes a conformational change in the factor which is similar to that caused by binding of GTP. Potentially, the effect of aurodox and kirromycin might be used to great advantage to stabilize EF-Tu·ribosome complexes for analysis by fluorescence techniques, thus opening up this approach to study the mechanism by which EF-Tu functions. All of these studies indicated above were carried out with EF-Tu isolated from *E. coli;* however, it is desirable to use labeled EF-Tu from *T. thermophilus* for experiments that utilize fluorescence as described in Sect. 3. Thus, we have investigated the effect of aurodox on EF-Tu from this source.

We conclude that unfortunately aurodox will be of little value for fluorescence studies in stabilizing various EF-Tu·ribosome complexes using labeled factor from *T. thermophilus.*

5.1 Effect of Aurodox on Enzymatic Binding of Phe-tRNA to Ribosomes

Poly(U)-directed synthesis of polyphenylalanine was inhibited by 50 % in the presence of 0.5 μM aurodox in either cell-free extracts from *T. thermophilus* or when synthesis was carried out with EF-Tu from this source, but with ribosomes, EF-G, and other components from *E. coli* (data not shown). Similar inhibition by kirromycin was observed using a cell-free system derived entirely from *E. coli.* (Wolf et al. 1974). In the presence of *T. thermophilus* EF-Tu and GTP, aurodox inhibited binding of Phe-tRNA to ribosomes slightly. With GDP no enzymatic binding of Phe-tRNA to ribosomes was observed in the presence or absence of aurodox (data not shown). These observations are in apparent contrast to the results from experiments in which *E. coli* EF-Tu was used (Chinali et al. 1977; Noort et al. 1982). They appear to reflect a difference in the effect of the antibiotic on the two factors and led us to investigate the mechanism by which the antibiotic inhibits peptide elongation in the presence of EF-Tu from the thermophile.

Fluorescence from the coumarin derivative of EF-Tu is particularly sensitive to changes in the local environment of the fluorophore. Changes in fluorescence from coumarin-EF-Tu caused by binding either aurodox or GDP are given in Table 4. Binding of GDP causes bathochromic and hypochromic changes of about 2 nm and 5.5 %, respectively. These changes were observed only with guanine nucleotides at the nucleotide concentrations used in the studies reported here. This appears to reflect the nucleotide specificity of EF-Tu. Aurodox alone caused a hypochromic change of 13.6 %.

Different guanine nucleotides caused qualitatively similar quenching; however, the magnitude of the changes varied. In the presence of aurodox, the largest effect, a 21 % decrease in fluorescence, was caused by GDP. Smaller changes were observed with GTP or the nonhydrolyzable analogues of GTP. This difference between the effect of GDP and GTP or its analogues cannot be explained by differences in dissociation constants of these nucleotides and the factor, because the nucleotides were present at relatively high concentrations ($> 30 \times K_d$). These results appear to reflect different conformational states of EF-Tu.

Table 4. Fluorescence changes of coumarin-EF-Tu caused by binding of nucleotides, Phe-tRNA or aurodox[a]

Addition	$\triangle F$ (%)	Q	Max emission wavelength (nm)	K_d nucleotide (nM)
None	0.0	0.72	468	
GDP	5.5	0.69	470	~ 10
GTP	6.0	0.68	470	~300
GDP-NP	4.0	0.70	470	2000
GDP-CP	4.3	0.70	469	3000
Phe-tRNA	0.0	0.72	468	
Aurodox	13.6	0.65	468	
GDP-NP + Phe-tRNA	4.0	0.69	470	
GDP + aurodox	21.0	0.60	470	~ 10
GTP + aurodox	19.0	0.60	471	~ 25
GDP-NP + aurodox	14.0	0.65	470	1000
GDP-CP + aurodox	16.8	0.63	470	1000

[a]For the quenching experiments, a 5 x 10 mm glass cuvette was used with 500 μl of a solution containing 20 mM Tris-HCl (pH 7.5), 10 mM magnesium acetate, 5 mM β-mercaptoethanol, and 100 nM coumarin-EF-Tu. Emission spectra were taken before and after addition of the following components to give the final concentration indicated: 100 μM guanine nucleotides, 20 μM aurodox, or 10 μM Phe-tRNA. When GTP was present, 1 mM phosphoenolpyruvate and 25 μg of pryruvate kinase were added. Fluorescence values were corrected for dilution and aurodox absorption at the excitation wavelength, 397 nm. $\triangle F$ is the decrease of fluorescence at the maximum emission wavelength; Q is the quantum yield. For determination of K_d, a 10 x 10 mm glass cuvette was used, and 1.7 ml mixtures containing 30 nM coumarin-EF-Tu were titrated with 50 μM GDP, 1 mM GDP-NP or GDP-CP

5.2 The Effect of Aurodox on the Interaction of Phe-tRNA with EF-Tu

Aminoacyl-tRNA is protected from nonenzymatic hydrolysis by the formation of a ternary complex with EF-Tu and GTP (Beres and Lucas-Lenard 1973). Thus, the rate of release of the amino acid can be used as a measure of the interaction between EF-Tu and aminoacyl-tRNA. The results of such experiments are shown in Figs. 2A and 2B which reflect the effect of guanine nucleotides and aurodox on the interaction of EF-Tu with Phe-tRNA. The rate of Phe-tRNA hydrolysis is dependent on the ionic strength. In the absence of EF-Tu, Phe-tRNA is more resistant to hydrolysis at higher concentrations of NH$_4$Cl. In the presence of GTP and a GTP-regenerating system, reduced hydrolysis of Phe-tRNA was observed over a wide range of NH$_4$Cl concentrations reflecting the formation of an EF-Tu·GTP·Phe-tRNA complex. This protection against hydrolysis is reduced by aurodox, particularly at higher concentrations of the salt. No protective effect was observed either in the presence or absence of aurodox if either GTP or EF-Tu were omitted from the reaction mixture or if GDP was substituted for GTP. The results of these experiments support the conclusion that binding of aurodox to EF-Tu interferes with the ability of the factor to interact with aminoacyl-tRNA. This interaction is promoted by GTP and inhibited by aurodox. The results give no indication that either in the absence or presence of GDP aurodox induces a conformation of EF-Tu from *T. thermophilus* similar to that induced by GTP that leads to binding and protection of aminoacyl-tRNA.

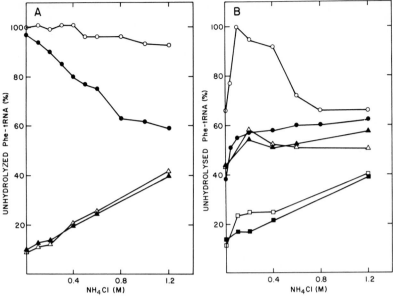

Fig. 2 A-B. Protection by EF-Tu of Phe-tRNA from hydrolysis. Reaction mixtures contained 0.8 nmol of EF-Tu, 0.15 nmol of [^{14}C]Phe-tRNA (25 Ci/mol) in 100 μl of 20 mM Tris-HCl (pH 7.5), 10 mM magnesium acetate, and 5 mM β-mercaptoethanol and the concentrations of NH$_4$Cl as indicated. *Closed symbols,* 20 μM aurodox; *open symbols,* no aurodox. **A** in the presence of 100 μM GTP, 2 mM phosphoenolpyruvate, and 5 μg of pyruvate kinase; with EF-Tu (*circles*), or EF-Tu replaced by bovine serum albumin (*triangles*). **B** in the presence of 100 μM nucleotides: GDP-NP (*circles*), GDP-CP (*triangles*); or GDP (*squares*). Reaction mixtures were incubated for 2 h at 37°C, then 90 μl of the mixtures was applied to Whatman 3MM filter discs and the amount of remaining [^{14}C]Phe-tRNA was determined

6. Structure and Function of S1

S1 is a highly elongated protein of about 230 Å (Giri and Subramanian 1977; Laughrea and Moore 1977), approximately equal to the longest dimensions of the *E. coli* ribosome. The amino terminal portion of the protein extending through amino acid residue 171 forms a discrete, trypsin-sensitive domain that functions in binding S1 to either ribosomes or the Qβ replicase (Subramanian 1983). Low angle X-ray scattering data of the trypsin-resistant portion from residue 172 to 557 at the C-terminus of the molecule indicate that it approximates a cylinder 200 Å in length (Subramanian, personal communication). Computer analysis of the amino acid sequence of this domain suggests a secondary structure that is rich in anti-parallel β sheets and β turns (Kimura et al. 1982). The proton nuclear magnetic resonance spectrum of S1 is consistent with an unusual degree of structural flexibility within the intact protein (Moore and Laughrea 1979), possibly resulting from flexibility between the N-terminal and C-terminal domains. The C-terminal domain contains the nucleic acid binding site (Subramanian 1983) and two cysteine residues at positions 292 and 349 which can be differentially labeled as described above. The more reactive thiol can be readily alkylated while the protein is bound to 30S subunits (Craven and Gupta 1970), but becomes resistant to alkylation when poly(U) or other polynucleotides are bound to the protein (Acharya and Moore 1973; Lipecky et al. 1977). The polynucleotide binding site appears to be at or very close to the two cysteine residues in the 3-dimensional structure (Odom et al. 1984). The dependence on S1 for polyphenylalanine synthesis is greatest with low concentrations of poly(U), but becomes progressively less with increasing amounts of the polynucleotide (Suryanarayana and Subramanian 1983). The molar ratio of S1 per ribosome required for optimum translation is approximately one (Miller et al. 1974) with inhibition of polyphenylalanine synthesis observed at higher S1 concentrations. In consideration of these results, recently Subramanian (1983) has proposed that S1 functions to sequester mRNA for translation by binding the nucleotide to the elongated C-terminal domain.

6.1 Binding of Poly(U) and ATA to S1

We have used fluorescence techniques to test the hypothesis indicated above (Odom et al. 1984). When poly(U) binds to S1 labeled with coumarin maleimide, the fluorescence emission maximum shifts to the red by about 3 nm with a 15 % decrease in the quantum yield. About the same amount of poly(U) is required to give the maximum shift in fluorescence and the maximum polyphenylalanine synthesis, suggesting that the nucleotide is bound simultaneously by S1 and at the decoding site of the ribosome during peptide elongation.

Other data indicate that a number of antibiotics that inhibit peptide synthesis on *E. coli* ribosomes have no effect detectable by fluorescence on the interaction of the coding nucleotide with labeled S1 free in solution or on ribosomes (Odom et al. 1984). However, the peptide initiation inhibitor, aurintricarboxylic acid, ATA, binds tightly to coumarin-labeled S1 free in solution or bound to ribosomes and almost completely quenches fluorescence from the probe (Odom et al. 1984). Quenching appears to be by energy transfer to the highly colored inhibitor. ATA was shown previously to block binding of poly(U) to S1 (Tal et al. 1972) and we demonstrated that prebinding of poly(U) to S1 inhibits subsequent interaction of the protein with ATA as judged by fluorescence quenching (Odom et al. 1984). These results appear to indicate that ATA binds to S1 at or near its polynucleotide binding site. They are consistent with the hypothesis that the effect of ATA on S1 may account for or contribute to its inhibitory effect on peptide initiation.

6.2 Distance Between the S1 Nucleotide Binding Site and the Ribosomal Decoding Site

The data indicated above suggest that a strand of mRNA being translated on an *E. coli* ribosome is bound simultaneously to S1 and at the decoding site where a codon of the polynucleotide base pairs with the anticodon of a cognate tRNA. Antibiotics that block various ribosomal steps of the peptide elongation cycle do not influence the interaction of the mRNA strand with S1 (Odom et al. 1984) suggesting that movement of the polynucleotide through S1 is not strictly coordinated with its movement through the other sites to which it is bound to the ribosome. These considerations introduce the question of the distance between the nucleotide binding site on S1 and the ribosomal decoding site. Energy transfer was measured between proflavin in the Y base position at the 3' side of the anticodon loop of yeast tRNA^{Phe} to S1 labeled in guanidine·HCl with eosin maleimide to give an eosin to S1 ratio of about 2. The reaction conditions for labeling S1, analysis of the labeled protein, and its reconstitution into 30S subunits were as described previously for fluorescein-S1 (Rychlik et al. 1983). The ratio of eosin-S1 bound to 30S subunits approached unity. Preparation of Phe-tRNA^{Phe} labeled at the Y base position with proflavin, Phe-tRNApf, was by the procedure of Wintermeyer and Zachau (1971) as described previously (Odom et al. 1978). Binding of Phe-tRNA$_{Pf}$ to the A site of 70S ribosomes containing eosin-S1 on the 30S subunit was carried out with EF-Tu and GTP, then energy transfer of 0.15 from proflavin to eosin was measured by the fluorescence lifetime method and analyzed as described previously (Robbins and Hardesty 1983). With an R_O of 49 Å determined for this proflavin to eosin pair, a distance of 65 Å with limits (Haas et al. 1978) of 59 Å and 72 Å was calculated, assuming energy transfer to only one, the closest, eosin residue. Probably, 65 Å is a minimum distance in that energy transfer to the second eosin residue would result in an underestimation of the true distance. These results appear to indicate that a considerable length of mRNA is required to span the gap between the polynucleotide binding site of S1 and the decoding site of the 30S subunit.

6.3 Flexibility of S1

The shape of S1 and its length relative to the size of the ribosome makes a study of its flexibility both free in solution and attached to ribosomes of particular interest in understanding its function. Time-resolved anisotropy of an AEDANS-S1 derivative was used to study this characteristic with the results shown in Fig. 3. The data of Fig. 3A are typical for labeled S1 free in solution and give rotational correlation times, τ_D, of

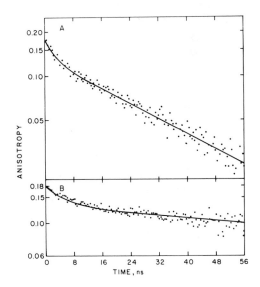

Fig. 3 A-B. Time-dependent decay of fluorescence anisotropy of S1 labeled with IAEDANS. The samples contained an $8 \times 10^{-7} M$ solution of labeled S1 in 10 mM Tris-HCl (pH 7.5), 10 mM magnesium acetate, 100 mM NH$_4$Cl, and 5 mM β-mercaptoethanol. **A** without 30S subunits; **B** in the presence of 1.5 x $10^{-6} M$ 30S subunits from which S1 had been removed

9 and 49 ns. The shorter time likely reflects local movement of the probe itself and/or an artifact generated by error in the value for scattered excitation light that was subtracted. It is not considered further. The longer value, 49 ns, apparently reflects depolarization caused by movement of the body of the S1 molecule. With a solution mol. w. of 61 159 (Schnier et al. 1982), a partial specific volume of 0.74 $cm^3 gm^{-1}$ (Subramanian 1983), and assuming a typical hydration of 0.32 g H_2O/g protein, the following conclusion can be reached. It can be calculated that if S1 were sperical, a τ_0 of 27 ns would be anticipated (Cantor and Tao 1971). However, S1 is a highly elongated protein. The ratio of its observed frictional coefficient to the frictional coefficient anticipated if the protein were sperical, f/f_0, is about 1.7 (Giri and Subramanian 1977; Yokota et al. 1979). Extrapolation of a plot of $\tau_D/\tau_{D,o}$ vs f/f_0 for a series of proteins in which these values are known (Cantor and Timasheff 1980) leads to an anticipated value for $\tau_D/\tau_{D,o}$ of about 2.0, corresponding to a τ_D value of 54 ns, if S1 in solution has the structure of a rigid prolate ellipsoid with an axial ratio of 12:1 (Subramanian 1983).

A similar calculation can be made for the trypsin-resistant domain from residue 172 to the C-terminal end of the molecule, assuming a rigid prolate ellipsoid 200 Å in length and an axial ratio of 10:1, as considered above. The frictional ratio, f/f_0, would be about 1.5 (Tanford 1961) with $\tau_D/\tau_{D,o}$ of about 1.85. Using this value and a mol. w. of 42 000, τ_D for the trypsin resistant fragment would be 35 ns. The data of Fig. 3B are consistent with a model of S1 free in solution in which a rigid, cylindrical, trypsin-resistant C-terminal domain is attached through a partially flexible hinge region to the N-terminal ribosome binding domain.

Experiments comparable to those described above with free S1 were carried out with AEDANS-S1 bound to 30S ribosomal subunits, with the results shown in Fig. 3B. The curve corresponds to τ_D values of 6 ns and 220 ns. Thus, binding of S1 to 30S subunits increases the longer τ_D value from about 50 ns to over 200 ns. The τ_D for the 30S subunit itself can be calculated by the procedure described above to be in excess of 500 ns. Very similar results to those shown in Fig. 3A and B were obtained using the pyrene maleimide derivative of S1 (Odom et al. 1984). Both of these sets of data appear to be incompatible with the results presented by Chu and Cantor (1979), who observed a rotational correlation time of 26 ns for free S1 under very similar experimental conditions with very little increase in τ_D of S1 attached to ribosomes (Chu et al. 1982).

The results shown in Fig. 3 indicate that binding of S1 to ribosomes causes a very significant increase in its rotational correlation time but not to the extent anticipated if some part of the protein was fixed rigidly to the 30S subunit. The results appear to be entirely consistent with a model in which S1 is bound to a 30S subunit only by the N-terminal domain with the movement of the rigid, rod-shaped C-terminal domain restricted by attachment at one end through a flexible hinge region. Presumably, mRNA in solution near the ribosome could be sequestered for translation by binding to the nucleotide binding site of the C-terminal domain. Also, S1 might play an important role in dissociating double-stranded regions of mRNA prior to translation. Thus, the data presented here appear to be consistent with the hypothesis of S1 function presented previously by Subramanian (1983).

Abbreviations

rRNA, ribosomal RNA; tRNA, transfer RNA; FTS, fluorescein 5'-thiosemicarbazide; ETS, eosin 5'-thiosemicarbazide; CPM, 3-(4-maleimidylphenyl)-4-methyl-7-diethyl-aminocoumarin; FM, fluorescein 5'-maleimide; PM, N-(3-pyrene)maleimide; CITC, 3-(4-isothiocyanatophenyl)-4-methyl-7-diethylaminocoumarin;FITC, fluorescein 5'-isothiocyanate; NMA, N-methylanthranilic acid hydrazide; Eos^H, 2-(N-[4'-eosin]-acetamidothio)-acethydrazide; Fluram, fluorescamine; PF, proflavine; IAEDANS, 1-(2-iodoacetamidoethylamino)naphthalene-5-sulfonic acid; Srd$_8$-C$_{13}$ cross-link, tRNA^{Phe} that has been photochemically cross-linked from 4-thiouridine in position 8 to cytidine in position 13; ATA, aurintricarboxylic acid.

Acknowledgement. We gratefully acknowledge the excellent technical assistance of M. Rodgers. We thank Drs. G. Kramer and S. Fullilove for critical discussion and review in preparing the manuscript, M. Powers for preparing the typescript, and F. Hoffman for photography and art work.

This work was supported in part by Grant PCM 81-12248 awarded by the National Science Foundation to Boyd Hardesty. Some fluorescence measurements were made at the Center for Fast Kinetics Research, which is supported by National Institutes of Health Grant RR-00886 from the Biotechnology Branch of the Division of Research Resources and by The University of Texas.

References

Acharya AS and Moore PB (1973) Reaction of ribosomal sulfhydryl groups with 5,5'-dithiobis(2-nitrobenzoic acid). J Mol Biol 76: 207-221

Arai K, Ota Y, Arai N, Nakamura S, Henneke C, Oshima T and Kaziro Y (1978a) Studies on poly-peptide-chain-elongation factors from an extreme thermophile, *Thermus thermophilus* HB8. 1. Purification and some properties. Eur J Biochem 92: 509-519

Arai K, Arai N, Nakamura S, Oshima T and Kaziro Y (1978b) Studies on polypeptide-chain-elonga-tion factors from an extreme thermophile, *Thermus thermophilus* HB8. 2. Catalytic properties. Eur J Biochem 92: 521-531

Beres L and Lucas-Lenard J (1973) Studies on the fluorescence of the Y base of yeast phenylalanine tRNA: Effect of pH, aminoacylation and interaction with elongation factor Tu. Biochemistry 12: 3998-4005

Blumenthal T, Douglass J and Smith D (1977) Conformational alteration of protein synthesis elonga-tion factor EF-Tu by EF-Ts and by kirromycin. Proc Natl Acad Sci USA 74: 3264-3267

Brauer D and Wittmann-Liebold B (1977) The primary structure of the initiation factor IF-3 from *E. coli*. FEBS Lett 79: 269-275

Cantor CR and Tao T (1971) Application of fluorescence techniques to the study of nucleic acids. In: Cantoni GL and Davies DR (eds) Procedures in Nucleic Acid Research, vol 2. Harper and Row, New York, pp 31-93

Cantor CR and Timasheff C (1980) In: Neurath H and Hill R (eds) The Proteins, vol. V. Academic, New York, p 527

Chinali G, Wolf H and Parmeggiani A (1977) Effect of kirromycin on elongation factor Tu. Location of the catalytic center for ribosome elongation-factor-Tu GTPase activity on the elongation factor. Eur J Biochem 75: 55-65

Chu YG and Cantor CR (1979) Segmental flexibility in *E. coli* ribosomal protein S1 as studied by fluorescence polarization. Nucleic Acids Res 6: 2363-2379

Chu YG, Cantor CR, Sawchyn I and Cole PE (1982) Segmental flexibility of ribosomal protein S1 bound to ribosomes and Qβ replicase. FEBS Lett 145: 203-207

Craven GR and Gupta V (1970) Three dimensional organization of the 30S ribosomal proteins from *E. coli*. I. Preliminary classification of the proteins. Proc Natl Acad Sci USA 67: 1329-1336

Dabbs E (1979) Selection for *E. coli* mutants with proteins missing from the ribosome. J Bacteriol 140: 734-737

Epe B, Steinhauser KG and Woolley P (1983) Theory of measurement of Förster-type energy transfer in macromolecules. Proc Natl Acad Sci USA 80: 2579-2583

Fasano O, Bruns W, Crechet J, Sander G and Parmeggiani A (1978) Modification of elongation-factor-Tu guanine-nucleotide interaction by kirromycin: A comparison with the effect of aminoacyl-tRNA and elongation factor Tu. Eur J Biochem 89: 557-565

Giri L and Subramanian AR (1977) Hydrodynamic properties of protein S1 from *E. coli* ribosome. FEBS Lett 81: 199-203

Haas E, Katchalski-Katzir E and Steinberg IZ (1978) Effect of the orientation of donor and acceptor on the probability of energy transfer involving electronic transitions of mixed polarization. Biochemistry 17: 5064-5070

Kimura M, Foulaki K, Subramanian AR and Wittmann-Liebold B (1982) Primary structure of *E. coli* ribosomal protein S1 and features of its functional domains. Eur J Biochem 123: 37-53

Laughrea M and Moore PB (1977) Physical properties of ribosomal protein S1 and its interaction with the 30S ribosomal subunit of *E. coli*. J Mol Biol 112: 399-421

Linde R, Khanh NQ and Gassen HG (1979) Purification of ribosomal protein S1 and physical tests of its homogeneity. Methods Enzymol 60: 417-426

Lipecky R, Kohlschein J and Gassen HG (1977) Complex formation between ribosomal protein S1, oligo- and polynucleotide chain length dependence and base specificity. Nucleic Acids Res 4: 3627-3642

Liu CM, Maehr H, Leach M, Liu M and Miller PA (1977) Biosynthesis of aurodox (antibiotic X-5108): Incorporation of [^{14}C]-labeled precursors into aurodox. J Antibiot Tokyo 30: 416-419

Liu TY (1977) The role of sulfur in proteins. In: Neurath H and Hill RL (eds) The Proteins, vol. III. Academic, New York, pp 240-402

Miller MJ, Niveleau A and Wahba AJ (1974) Initiation of synthetic and natural messenger translation: Purification and properties of a protein isolated from *E. coli* MRE 600 ribosomes. J Biol Chem 249: 3803-3807

Moore PB (1980) Scattering studies of the three-dimensional organization of the *E. coli* ribosome. In: Chambliss G, Craven GR, Davies J, Davis K, Kahan L and Nomura M (eds) Ribosomes: structure, function and genetics. University Park Press, Baltimore, Maryland, pp 111-133

Moore PB and Laughrea M (1979) The conformational properties of ribosomal protein S1. Nucleic Acids Res 6: 2355-2361

Nakamura S, Ohta S, Arai K, Arai N, Oshima T and Kaziro Y (1978) Studies on polypeptide-chain-elongation factors from an extreme thermophile, *Thermus thermophilus* HB8. 3. Molecular properties. Eur J Biochem 92: 533-543

Nierhaus KH, Lietke R, May RP et al. (1983) Shape determinations of ribosomal proteins *in situ*. Proc Natl Acad Sci USA 80: 2889-2893

Nomura M and Held WA (1974) Reconstitution of ribosomes: Studies of ribosome structure, function and assembly. In: Nomura M, Tissieres A and Lengyel P (eds) Ribosomes. Cold Spring Harbor, New York, pp 193-223

Noort JM, Duisterwinkel FJ, Jonak J, Sedlacek J, Kraal B and Bosch L (1982) The elongation factor Tu·kirromycin complex has two binding sites for tRNA molecules. EMBO J 1: 1199-1205

Odom OW, Craig BB and Hardesty B (1978) The conformation of the anticodon loop of yeast tRNA^{Phe} in solution and on ribosomes. Biopolymers 17: 2909-2931

Odom OW, Deng HY, Subramanian AR and Hardesty B (1984) Relaxation time, interthiol distance, and mechanism of action of ribosomal protein S1. Arch Biochem Biophys 230: 178-193

Odom OW, Jr, Robbins DJ, Lynch J, Dottavio-Martin D, Kramer G and Hardesty B (1980) Distances between 3' ends of ribosomal ribonucleic acids reassembled into *E. coli* ribosomes. Biochemistry 19: 5947-5954

Ofengand J (1980) The topography of tRNA binding sites on the ribosome. In: Chambliss G, Craven GR, Davies J, Davis K, Kahan L and Nomura M (eds) Ribosomes: structure, function and genetics. University Park Press, Baltimore, Maryland, pp 497-530

Ofengand J, Lin FL, Hsu L and Boublik M (1981) Three-dimensional arrangement of tRNA at the donor and acceptor sites of the *E. coli* ribosome. In: Siddiqui M, Krauskopf M and Weissbach H (eds) Molecular approaches to gene expression and protein structure. Academic, New York, pp 1-31

Parmeggiani A and Sander G (1980) In: Sammes PG (ed) Topics in antibiotic chemistry, vol 5. Ellis Harwood, Chichester, pp 160-221

Robbins D and Hardesty B (to be published 1984) Comparison of ribosomal entry and acceptor tRNA binding sites on *E. coli* 70S ribosomes: Fluorescence energy transfer measurements from Phe-tRNA^{Phe} to the 3' end of 16S RNA. Biochemistry

Robbins DR, Odom OW Jr, Lynch J, Kramer G, Hardesty B, Liou R and Ofengand J (1981) Position of tRNA on *E. coli* ribosomes: Distance from the 3' end of 16S RNA to three points on phenyl-alanine-accepting tRNA in the donor site of 70S ribosomes. Biochemistry 20: 5301-5309

Rychlik W, Odom OW and Hardesty B (1983) Localization of the elongation factor Tu binding site on *E. coli* ribosomes. Biochemistry 22: 85-93

Schnier J, Kimura M, Foulaki K, Subramanian AR, Isono K and Wittmann-Liebold B (1982) Primary structure of *E. coli* ribosomal protein S1 and of its gene *rpsA*. Proc Natl Acad Sci USA 79: 1008-1011

Steitz JA (1980) RNA·RNA interactions during polypeptide chain initiation. In: Chambliss G, Craven GR, Davies J, Davis K, Kahan L and Nomura M (eds) Ribosomes: structure, function and genetics. University Park Press, Baltimore, Maryland, pp 479-496

Stöffler G, Bald R, Kastner B, Lührmann R, Stöffler-Meilicke M, Tischendorf G and Tesche B (1980) Structural organization of the *E. coli* ribosome and localization of functional domains. In: Chambliss G, Craven GR, Davies J, Davis K, Kahan L and Nomura M (eds) Ribosomes: structure, function and genetics. University Park Press, Baltimore, Maryland, pp 171-206

Stöffler-Meilicke M, Stöffler G, Odom OW, Zinn A, Kramer G and Hardesty B (1981) Localization of 3' ends of 5S and 23S rRNAs in reconstituted subunits of *E. coli* ribosomes. Proc Natl Acad Sci USA 9: 5538-5542

Subramanian AR, Rienhardt P, Kimura M and Suryanarayana T (1981) Fragments of ribosomal protein S1 and its mutant form m1-S1: Localization of nucleic acid-binding domain in the middle region of S1. Eur J Biochem 119: 245-249

Subramanian AR (1983) Structure and functions of ribosomal protein S1. Progr Nucleic Acid Res Mol Biol 28: 101-142

Suryanarayana T and Subramanian AR (1983) An essential function of ribosomal protein S1 in messenger ribonucleic acid translation. Biochemistry 22: 2715-2719

Tal M, Aviram M, Kanarek A and Weiss A (1972) Polyuridylic acid binding and translating by *E. coli* ribosomes: Stimulation by protein I, inhibition by aurintricarboxylic acid. Biochem Biophys Acta 281: 381-392

Tanford C (1961) Physical Chemistry of Macromolecules. Wiley, New York, p 326

Traut RR, Lambert JM, Boileau G and Kenny JW (1980) In: Chambliss G, Craven GR, Davies J, Davis K, Kahan L and Nomura M (eds) Ribosomes: structure, function and genetics. University Park Press, Baltimore, Maryland, pp 89-110

Wintermeyer W and Zachau H (1971) Replacement of Y base, dihydrouracil, and 7-methylguanine in tRNA by artificial odd bases. FEBS Lett 18: 214-218

Wittmann HG (1983) Architecture of prokaryotic ribosomes. Ann Rev Biochem 52: 35-65

Wittmann H, Littlechild J and Wittmann-Liebold B (1980) Structure of ribosomal proteins. In: Chambliss G, Craven GR, Davies J, Davis K, Kahan L and Nomura M (eds) Ribosomes: structure, function and genetics. University Park Press, Baltimore, Maryland, pp 51-88

Wolf H, Chinali G and Parmeggiani A (1974) Kirromycin, an inhibitor of protein biosynthesis that acts on elongation factor Tu. Proc Natl Acad Sci USA 71: 4910-4914

Yokota T, Arai K and Kaziro Y (1979) Studies on 30S ribosomal protein S1 from *E. coli*. I. Purification and physicochemical properties. J Biochem (Tokyo) 86: 1725-1737

Ribosomal Components of the Peptidyl Transferase Center

E. KÜCHLER, G. STEINER, and A.BARTA[1]

1. Introduction

The past years have seen tremendous progress in our knowledge of the structures of ribosomal proteins and RNA (for review articles see Wittmann 1983; Noller 1984, Wollenzien et al. 1984). Yet our understanding of the ribosomal mechanisms has remained rather fragmentary. The extremely complex interplay of the various factors involved in initiation, elongation, and termination, the interactions of mRNA and tRNA with the various ribosomal components, and finally the molecular processes involved in transpeptidation and translocation still represent a "terra incognita" in our concept of protein biosynthesis. However, techniques have been developed in the last few years which today allow us to focus on at least some of the reactions taking place on an actively translating ribosome (Spirin and Serdyuk, 1984). Chemical methods such as affinity and photoaffinity labelling techniques and cross-linking with bifunctional reagents or UV light have been employed to detect functional domains on the ribosome. In particular affinity and photoaffinity reactions have been used successfully to study the ribosomal binding sites for tRNA, mRNA, and antibiotics. The literature on this subject is too extensive to be discussed here in any detail. For summaries on the work performed the reader is referred to several review articles (Pellegrini and Cantor 1977; Küchler 1978; Cantor 1979; Küchler and Ofengand 1979; Johnson 1979; Ofengand 1980).

The site on the ribosome, where transpeptidation occurs, is an integral part of the large ribosomal subunit and is usually referred to as the peptidyl transferase center. Studies in the past have concentrated primarily on ribosomal proteins, for the peptidyl transferase had usually been considered to be a catalytic center similar to the active site of an enzyme. Indeed, when the peptidyl site (P-site) is occupied by derivatives of radioactively labelled aminoacyl-tRNA, such as bromoacetyl- and p-nitrophenoxy-carbonyl-Phe-tRNA, several proteins of *E. coli* ribosomes, such as L2, L15, and L27 have been found to be labelled (Pellegrini et al. 1972; Czernilofsky and Küchler 1972; Oen et al. 1973; Czernilofsky et al. 1974; Pellegrini et al. 1974). The criteria applied to demonstrate specificity of the reaction were the dependence of the labelling on the presence of poly(U) as messenger RNA as well as the inhibition by the addition of puromycin, an antibiotic known to act at the ribosomal acceptor site (A-site) by simulating the free end of an aminoacyl-tRNA. The radioactively labelled aminoacyl moiety carrying the affinity probe is transferred from the tRNA at the P-site onto puromycin

[1] Institut für Biochemie, Universität Wien, 1090 Vienna, Währinger Straße 17, Austria

Mechanisms of Protein Synthesis
ed. by Bermek
© Springer-Verlag Berlin Heidelberg 1984

and is released from the ribosome, thus reducing incorporation into ribosomal proteins. In other experiments it was shown that IF2 recognizes the corresponding affinity label derivative of initiator tRNA by demonstrating that this factor specifically stimulates affinity labelling of the ribosomal proteins (Hauptmann et al. 1974; Sopori et al. 1974; Hauptmann et al. unpublished results). A more direct way of proving that the affinity labelling occurs at the peptidyl transferase center was first described by Pellegrini et al. (1972). In these experiments the affinity labelling was carried out with a reactive derivative of nonradioactive aminoacyl-tRNA and radioactively labelled aminoacyl-tRNA was added subsequently. The radioactive amino acid could only be linked covalently via peptide bond formation, thus proving that the affinity labelling of the ribosomal proteins has indeed occured at the peptidyl transferase center. Affinity labelling experiments have also been carried out with reactive derivatives of various antibiotics acting on the peptidyl transferase (for a review see Cooperman 1980). Although these experiments were very successful in terms of topographically defining a functional domain on the ribosome, and in fact have been independently confirmed by immune electron microscopy (Stöffler et al. 1980; Lake 1980), the original goal of identifying a single protein as the peptidyl transferase has not been fulfilled. It has also proved difficult to obtain a clear correlation between the results of the affinity labelling experiments and studies on reconstitution of peptidyl transferase activity by reassembly of 50S subunits from isolated ribosomal components. Yet most of the proteins labelled in the affinity reactions have been shown to be either essential for or strongly stimulatory in reconstitution of peptidyl transferase activity (Nierhaus 1982; Fahnestock 1975).

2. 23S RNA at the Peptidyl Transferase Center

It has been generally observed that affinity and photoaffinity probes capable of reacting both with proteins and RNA preferentially label 23S RNA (Küchler and Ofengand 1979; Johnson 1979; Ofengand 1980). In fact the first affinity derivative of aminoacyl-tRNA (chlorambucylyl-Phe-tRNA) reported to react with the ribosome was found to label 23S RNA (Bochkareva et al. 1971). Unfortunately, at that time the identification of the site of labelling on a molecule as large as 23S RNA was virtually impossible. Recently, however, the complete sequence of the 2904 nucleotides of E. coli 23S RNA has been determined (Brosius et al. 1980) and techniques have been developed which permit the identification of reaction sites even on a nucleic acid as complex as 23S RNA.

In this paper we wish to discuss studies on the ribosomal peptidyl transferase center carried out with the photoaffinity label 3-(4'-benzoylphenyl)propionyl-Phe-tRNA (BP-Phe-tRNA). Aromatic ketone derivatives are suitable photoaffinity reagents for ribosomes, because they are activated by UV irradiation at 320 nm, a procedure which does not result in ribosomal inactivation (Küchler and Barta 1977). Photoactivation results in the transition of a lone electron pair ($n \rightarrow \pi^*$ transition) which leads to an intermediate triplet state. The energy of this activated state is sufficient for a reaction with peptides as well as nucleotides as has been shown in model experiments (Galardy et al. 1973; Barta and Küchler 1983). The activation energy is, however, too low to allow reaction with water molecules, so that loss due to side reactions should be negligible.

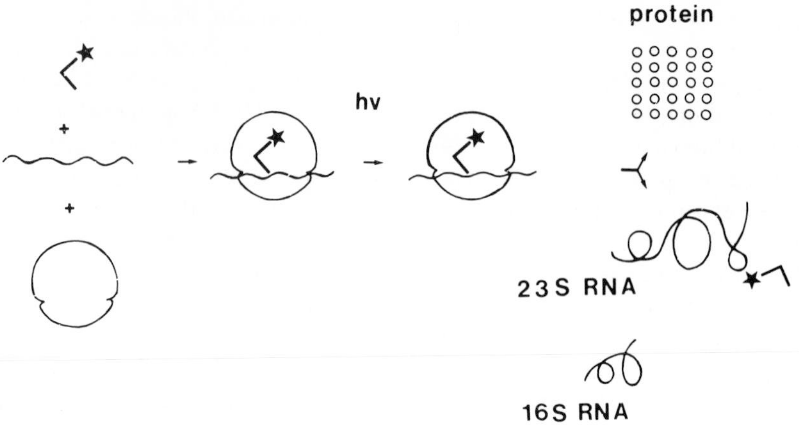

Fig. 1. Photoaffinity labelling on poly(U)-primed ribosomal complexes. The *asterisk* represents the photoreactive probe attached to the radioactive Phe-tRNA molecule

The design of such a photoaffinity labelling experiment is shown in Fig. 1. The asterisk on the L-shaped tRNA indicates the 3-(4'-benzoylphenyl)propionyl residue (BP) attached to the radioactive Phe moiety. The complex is formed using 70S ribosomes (tight couples) and poly(U) as messenger RNA. The incubation mixture is then cooled and irradiated at 320 nm using a high pressure mercury lamp. Upon photoreaction the BP residue of the BP-Phe-tRNA can attach to residues located at the peptidyl transferase center with the radioactive Phe serving as a tracer. In order to identify the site of reaction, ribosomes were denatured and proteins as well as ribosomal RNA were analyzed for radioactivity. Under these conditions incorporation was observed almost exclusively in 23S RNA (Barta et al. 1975). The following control experiments serve to demonstrate the specificity of the reaction. Photoaffinity labelling of 23S RNA was shown to be dependent on the presence of poly(U) as messenger RNA and was inhibited upon preincubation with puromycin before irradiation, indicating that the labelling occurred at the ribosomal P-site. As expected there was no incorporation when the sample was not irradiated. Further proof for photoaffinity labelling at the peptidyl transferase center was provided by a transpeptidation assay. In this experiment the photoaffinity reaction was carried out with nonradioactive BP-Phe-tRNA, [³H]Phe-tRNA being added subsequently. [³H]Phe was indeed bound covalently via transpeptidation to the BP-Phe residue already attached to the 23S RNA. In a parallel experiment the order was reversed. Following complex formation with nonradioactive BP-Phe-tRNA, [³H]Phe-tRNA was added to allow peptide bond formation and irradiation was carried out subsequently. A similar amount of radioactivity was recovered in 23S RNA in both experiments indicating that transpeptidation is extremely efficient even after photocrosslinking has occurred (Barta and Kuechler 1983).

As a first step to localize the area of the site of reaction on 23S RNA, degradation studies were performed with matrix-bound ribonuclease A. As shown in Fig. 2, the enzyme cuts the 23S RNA into a 13S and an 18S fragment comprising one-third from the 5' end and two-thirds from the 3' end, respectively. Photocrosslinking of the 18S

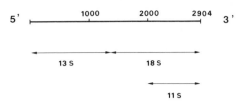

Fig. 2. Fragmentation of 23S RNA by matrix-bound ribonuclease A. The radioactive photo-affinity probe was found to be covalently attached to the 18S and 11S fragments, respectively

fragment was observed (Barta et al. 1975). A similar result was obtained with a different label (Sonenberg et al. 1975; 1977). Further fragmentation results in an 11S fragment representing sequences at the 3' end of 23S RNA. The photoaffinity label was found to be attached to this 11S RNA fragment indicating that the site of reaction lies within a region of about one thousand nucleotides from the 3' terminus (Barta and Kuechler, 1983).

Identification of the site of the photoreaction was performed by taking advantage of the plasmid pKK123 which contains an insert of the rrnB operon covering the region of 23S RNA from the base 843 to 2904 as shown in Fig. 3 (Barta et al. 1984). Upon

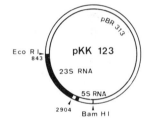

Fig. 3. Plasmid pKK 123 containing an insert of ribosomal DNA representing part of the rrnB operon

digestion with the restriction enzymes *EcoRI* and *Bam HI* this insert can be isolated. For the localization of the area labelled by BP-Phe-tRNA, the *EcoRI – Bam HI* fragment was cleaved with various restriction enzymes and the fragments were separated by agarose gel electrophoresis. The gel was then blotted onto nitrocellulose according to the technique of Southern and the filter hybridized with 23S RNA labelled with BP-[^{3}H]Phe-tRNA. The filter was subsequently treated with ribonuclease in order to remove non-hybridized parts of the 23S RNA molecule. The scheme of the experiment is shown in Fig. 4, the sizes and the positions of the cleavage sites are indicated. The following fragments gave positive hybridization signals with this probe (Fig. 4): a *Hinf I* fragment of 1821 base pairs starting at position 2004 reaching beyond the end of the 23S RNA gene; either one of two *Hae III* fragments of 431 and 424 base pairs of length situated next to each other at positions 1983 to 2414 and 2415 to 2838, respectively; a *Hpa II* fragment of 224 base pairs covering the region 2442 to 2666; and finally a 264 base pair *Cfo I* piece corresponding to bases 2361 through 2625. The area of the photocrosslink is defined by the two closest restriction cleavage sites, i.e., between bases 2442 and 2625, as shown in Fig. 4 (Barta et al. 1984).

The exact position of the site of the photoreaction was determined by extension of radioactively labelled primers in a reverse transcription experiment. Primers were

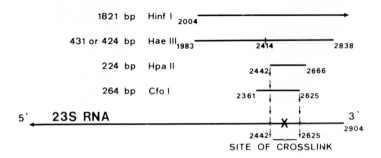

Fig. 4. Hybridization of rDNA fragments to photoaffinity 23S RNA. Location of the restriction fragments used for the identification of the site of cross-link

obtained by isolating restriction fragments located to the 3' side of the area identified as the site of the crosslink in the previous hybridization experiment. The restriction fragments were ^{32}P-labelled at their 5' termini and the strands separated by gel electrophoresis. The strand complementary to 23S RNA was then used to prime the reverse transcription reaction. The experiment is described in Fig. 5. Reverse transcription will proceed along the RNA from 3' to 5' as indicated by the arrow. If, however, the enzyme encounters a modified nucleotide it will either slow down or stop completely depending on the particular modification. As shown by Youvan and Hearst (1979; 1981), even a methyl group substitution on a base, such as a m^2G, will act as a kinetic barrier, whereas a large modification will cause the enzyme to stop transcription completely. The newly synthesized DNA strand is terminated one nucleotide before the modified base. Since the attachment of a BP-Phe-tRNA molecule constitutes a drastic modification, one would expect primer extension to stop just before the photoaffinity-labelled base. This experiment utilized two different primers, one being complementary to a region between the bases 2715 and 2839, the other corresponding to bases 2607 to 2663. Independent of which primer was used, cDNA chain extension by reverse transcriptase was found to stop at residues U2585 and U2586. Assuming that the results of Youvan and Hearst (1981) also hold true for the photoaffinity system, the residues labelled in the photoaffinity reaction should be U2584 and U2585, respectively. On the basis of

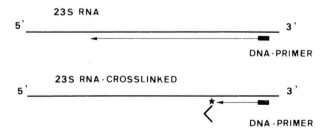

Fig. 5. Scheme of the primer extension experiment. The direction of reverse transcription is indicated by the *arrows*. Other symbols are as in Fig. 1

this experiment alone it cannot be excluded that in this particular situation the reverse transcriptase stops earlier and/or that termination occurs at two adjacent nucleotides even though only a single base has been modified. This question requires further experimentation, although independent evidence has shown the photoaffinity reaction to occur with a pyrimidine residue (Barta and Küchler 1983).

More important than the nature of the modified base itself is the region of 23S RNA in which the photoaffinity reaction has occurred. The residues U2584 and U2585 are located in the central loop of domain V (Fig. 6), a sequence which is highly conserved in evolution (Noller et al. 1981; Branlant et al. 1981). Both its structural characteristics as well as some of the nucleotide sequences are constant in large ribosomal RNAs from sources as different as bacteria, yeast, and human mitochondria, chloroplasts, and lower and higher eukaryotes (for a review see Noller 1984). Furthermore, as shown in Fig. 6 mutations in mitochondria leading to resistance to the antibiotic chloramphenicol and one mutation conferring resistance to erythromycin are located within the same loop region (Dujon 1980; Kearsay and Craig 1981; Sor and Fukuhara 1982). The mode of action of chloramphenicol is well-defined; it is known to act directly at the peptidyl transferase center inhibiting the transpeptidation reaction (Monro and Vazquez 1967). There are no comparable mutants known in *E. coli,* but this is presumably due to the fact that there are several genes coding for ribosomal RNA making it very difficult to obtain such mutants. Mitochondria, in contrast, have only one set of genes for rRNA. Because of the characteristic loop structure and because of the conservation of nucleic acid sequence, the central loop of domain V can be clearly distinguished within sequences from the different ribosomal RNAs. For this reason, it is generally assumed that the results on the antibiotic-resistant mutants in mitochondria are also relevant for bacterial ribosomes (Noller 1984).

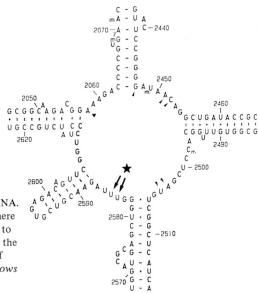

Fig. 6. Central loop of domain V of 23S RNA. The small *arrowheads* point at the sites where base changes occur as a result of mutation to chloramphenicol resistance, the *triangle* at the site of erythromycin resistance. The site of photoaffinity labelling is shown by the *arrows* and the *asterisk*

3. Conclusion

It appears to be highly significant that two independent approaches, i.e., photoaffinity labelling and the study of mutants conferring resistance to the antibiotics chloramphenicol and erythromycin, indicate the same region of 23S RNA to be a constituent of the peptidyl transferase center. This immediately raises the question of the role of ribosomal RNA in transpeptidation. There is ample evidence that the interaction between the CCA terminus of tRNA and the ribosome is highly specific (Monro et al. 1968). In a ribosomal complex these two C residues are strongly protected from chemical modification (Peattie and Herr 1981). The most straightforward interpretation is that the CCA terminus interacts with a complementary nucleotide sequence in the ribosomal RNA (Noller et al. 1981). As the CCA terminus is present in all tRNAs a highly conserved complementary sequence should occur in 23S RNA unless nonclassical Watson-Crick pairs are postulated. Within the same loop of domain V there are two G residues present at position 2607 and 2608. According to the secondary structure model of the 23S RNA these two G residues seem to be close enough to interact with the two C residues of the CCA and at the same time allow photoaffinity labelling of the U residues at position 2584 and 2585 (Noller 1984). This GG sequence is highly conserved in bacteria, chloroplasts, and mitochondria, but appears as AG in cytoplasmic ribosomes of eukaryotes. It is unlikely that such a high degree of variation could be tolerated. Outside the central loop region of domain V there is a UGG_{809}, a UGG_{2251}, and a UGG_{2686} conserved in bacteria. However, UGG_{2251} is changed into CGG in mitochondrial RNA of Paramecium and UGG_{2686} is not present at all in mitochondria of mammals. Only UGG_{809} is conserved in the 16 different 23S-like RNAs sequenced so far (Noller 1984). It should be borne in mind, though, that tertiary interactions could bring RNA regions together which appear very distant in the primary and secondary structures (see Wollenzien et al. 1984).

For many years ribosomal RNA was considered simply to be a kind of framework holding the various proteins in the proper positions and configurations. This general picture changed dramatically with the discovery that the 3' terminus of the 16S RNA was actively involved in initiation by interacting with sequences on the 5' side of the initiation codon on the mRNA (Shine and Dalgarno 1974; Steitz and Jakes 1975). Peptidyl transferase has so far been discussed mostly in terms of ribosomal proteins catalyzing a sequence of reactions which is basically the inverse of proteolysis (Nierhaus 1982). On the other hand studies on photoaffinity and affinity labelling together with the work on resistance mutants underline the importance of the ribosomal RNA as a constituent of the peptidyl transferase center. Description of the roles of the individual ribosomal components and elucidation of the molecular mechanism of the peptidyl transferase reaction presents a challenging subject for future experimentation.

*Acknowledgements.*This work was supported by a grant from the Austrian "Fonds zur Förderung der wissenschaftlichen Forschung". We thank H. Noller for sending a preprint and for helpful discussions, A. Gupta for technical assistance, T. Skern for critical reading, and B. Gamperl for typing of the manuscript.

References

Barta A, Küchler E (1983) Part of the 23S RNA located in the 11S RNA fragment in a constituent of the ribosomal peptidyl transferase center. FEBS Lett 163: 319-323

Barta A, Küchler E, Branlant C, SriWidada J, Krol A, Ebel JP (1975) Photoaffinity labelling of 23S RNA at the donor-site of the *Escherichia coli* ribosome. FEBS Lett 56: 170-174

Barta A, Steiner G, Brosius J, Noller HF, Küchler E (1984) Identification of a site on 23S ribosomal RNA located at the peptidyl transferase center. Proc Natl Acad Sci USA 81: 3607-3611

Bochkareva ES, Budker VG, Girshovich AS, Knorre DG, Teplova NM (1971) An approach to specific labeling of ribosomes in the region of peptidyl transferase center using N-acylaminoacyl-tRNA with an active alkylating grouping. FEBS Lett 19: 121-124

Branlant C, Krol A, Machatt MA, Pouyet J, Ebel JP, Edwards K, Kössel H (1981) Primary and secondary structures of *Escherichia coli* MRE 600 23S ribosomal RNA. Comparison with models of secondary structure for maize chloroplast 23S rRNA and for large portions of mouse and human 16S mitochondrial rRNAs. Nucleic Acids Res 9: 4303-4324

Brosius J, Dull TJ, Noller HF (1980) Complete nucleotide sequence of a 23S ribosomal RNA gene from *Escherichia coli*. Proc Natl Acad Sci USA 77: 201-204

Cantor CR (1979) tRNA – ribosome interactions. In: Schimmel PR, Söll D, Abelson JN (eds) Transfer RNA: structure, properties and recognition. Cold Spring Harbor Laboratory, New York, p 363

Cooperman BS (1980) Functional sites on the *E. coli* ribosome as defined by affinity labelling. In: Chambliss G, Craven GR, Davies J, Davis K, Kahan L, Nomura M (eds) Ribosomes – structure, function and genetics. University Park Press, Baltimore, p 531

Czernilofsky AP, Collatz E, Stöffler G, Küchler E (1974) Proteins at the tRNA binding sites of *Escherichia coli* ribosomes Proc Natl Acad Sci USA 71: 230-234

Czernilofsky AP, Küchler E (1972) Affinity label for the tRNA binding site on the *Escherichia coli* ribosome. Biochem Biophys Acta 272: 667-671

Fahnestock S (1975) Evidence of the involvement of a 50S ribosomal protein in several active sites. Biochemistry 14: 5321-5327

Galardy RE, Craig LC, Printz MP (1973) Benzophenone triplet: a new photochemical probe of biological ligand – receptor interactions. Nature New Biol 242: 127-128

Dujon B (1980) Sequence of the intron and flanking exons of the mitochondrial 21S rRNA gene of yeast strains having different alleles at the W and rib-1 Loci. Cell 20: 185-197

Hauptmann R, Czernilofsky AP, Voorma HO, Stöffler G, Küchler E (1974) Identification of a protein at the ribosomal donor-site by affinity labeling. Biochem Biophys Res Comm 56: 331-337

Johnson AE (1979) Analogs of lysyl-tRNA as probes of ribosome and elongation factor Tu structure and function. In: Schimmel PR, Söll D, Abelson JN (eds) Transfer RNA: structure, properties and recognition. Cold Spring Harbor Laboratory, New York, p 487

Kearsey SE, Craig IW (1981) Altered ribosomal RNA genes in mitochondria from mammalian cells with chloramphenicol resistance. Nature (Lond) 290: 607-608

Küchler E (1978) Affinity labels for tRNA and mRNA binding sites on ribosomes. In: Sundaram PV, Eckstein F (eds) Theory and practise in affinity techniques. Academic, London, p 151

Küchler E, Barta A (1977) Aromatic ketone derivatives of aminoacyl-tRNA as photoaffinity labels for ribosomes. In: Jakoby WB, Wilchek M (eds) Methods in enzymology, vol XLVI. Academic, New York, p 676

Küchler E, Ofengand J (1979) Affinity labeling of tRNA binding sites on ribosomes. In: Schimmel PR, Söll D, Abelson JN (eds) Transfer RNA: structure, properties and recognition. Cold Spring Harbor Laboratory, New York, p 413

Lake JA (1980) Ribosome structure and functional sites. In: Chambliss G, Craven GR, Davies J, Davis K, Kahan L, Nomura M (eds) Ribosomes – structure, function and genetics. University Park Press, Baltimore, p 207

Monro RE, Cerna J, Marcker KA (1968) Ribosome-catalyzed peptidyl transfer: substrate specificity at the P-site. Proc Natl Acad Sci USA 61: 1042-1049

Monro RE, Vazquez D (1967) Ribosome-catalyzed peptidyl transfer: effects of some inhibitors of protein synthesis. J Mol Biol 28: 161-165

Nierhaus KH (1982) Structure, assembly and function of ribosomes. Curr Top Microbiol Immunol 97: 81-155

Noller HF (1984) Structure of ribosomal RNA. Ann Rev Biochem 53; 119-162

Noller HF, Kop JA, Wheaton V, Brosius J, Gutell RR, Kopylov AM, Dohme F, Herr W, Stahl DA, Gupta R, Woese CR (1981) Secondary structure model for 23S ribosomal RNA. Nucleic Acids Res 9: 6167-6189

Oen H, Pellegrini M, Eilat D, Cantor CR (1973) Identification of 50S proteins at the peptidyl-tRNA binding site of *Escherichia coli* ribosomes. Proc Natl Acad Sci USA 70: 2799-2803

Ofengand J (1980) The topography of tRNA binding sites on the ribosome. In: Chambliss G, Craven GR, Davies J, Davis K, Kahan L, Nomura M (eds) Ribosomes – structure, function and genetics. University Park Press, Baltimore, p 497

Peattie DA, Herr W (1981) Chemical probing of the tRNA – ribosome complex. Proc Natl Acad Sci 78: 2273-2277

Pellegrini M, Cantor CR (1977) Affinity labeling of ribosomes. In: Weissbach H, Pestka S (eds) Molecular mechanism of protein biosynthesis. Academic, London, p 203

Pellegrini M, Oen H, Cantor CR (1972) Covalent attachment of a peptidyl-transfer RNA analog to the 50S subunit of *Escherichia coli* ribosomes. Proc Natl Acad Sci USA 69: 837-841

Pellegrini M, Oen H, Eilat D, Cantor CR (1974) The mechanism of covalent reaction of bromo-acetyl-phenylalanyl-transfer RNA with the peptidyl-transfer RNA binding site of the *Escherichia coli* ribosomes. J Mol Biol 88: 809-829

Shine J, Dalgarno L (1974) The 3'-terminal sequence of *Escherichia coli* 16S ribosomal RNA: complementarity to nonsense triplets and ribosome binding sites. Proc Natl Acad Sci USA 71: 1342-1346

Sonenberg N, Wilchek M, Zamir A (1975) Identification of a region in 23S rRNA located at the peptidyl transferase center. Proc Natl Acad Sci USA 72: 4332-4336

Sonenberg N, Wilchek M, Zamir A (1977) Mapping of 23S rRNA at the ribosomal peptidyl-trans-ferase center by photoaffinity labeling. Eur J Biochem 77: 217-222

Sopori M, Pellegrini M, Lengyel P, Cantor CR (1974) Affinity labeling of *E. coli* ribosomal prote-ins with an analog of the natural initiator tRNA. Biochemistry 13: 5432-5439

Sor F, Fukuhara H (1982) Identification of two erythromycin resistance mutations on the mito-chondrial gene coding for the large ribosomal RNA in yeast. Nucleic Acids Res 10: 6571-6577

Spirin AS, Serdyuk IN (1984) this volume

Steitz JA, Jakes K (1975) How ribosomes select initiator regions in mRNA: base pair formation be-tween the 3'terminus of 16S rRNA and the mRNA during initiation of protein-synthesis in *Escherichia coli*. Proc Natl Acad Sci USA 72: 4734-4738

Stöffler G, Bald R, Kastner B, Lührmann R, Stöffler-Meilicke M, Tischendorf G, Tesche B (1980) Structural organization of the *Escherichia coli* ribosome and localization of functional domains. In: Chambliss G, Graven GR, Davies J, Davis K, Kahan L, Nomura M (eds) Ribosomes – struc-ture, function and genetics. University Park Press, Baltimore, p 171

Wittmann HG (1983) Architecture of prokaryotic ribosomes. Ann Rev Biochem 52: 35-65

Wollenzien PL, Hui CF, Kang C, Murphy F, Cantor CR (1984) this volume

Youvan DC, Hearst JE (1979) Reverse transcriptase pauses at N^2-methylguanine during *in vitro* transcription of *Escherichia coli* 16S ribosomal RNA. Proc Natl Acad Sci USA 76: 3751-3754

Youvan DC, Hearst JE (1981) A sequence from *Drosophila melanogaster* 18S rRNA bearing the conserved hypermodified nucleoside amψ: analysis by reverse transcription and high-performance liquid chromatography. Nucleic Acids Res 9: 1723-1741

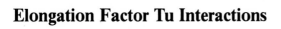

Elongation Factor Tu Interactions

Interaction Between Elongation Factor Tu and Aminoacyl-tRNA

A. Parmeggiani[1], G. Parlato[1,2], J. Guesnet[1], D. Picone[1,2], R. Pizzano[1,2], and O. Fasano[1,2]

1. Introduction

Among the various factors needed for protein biosynthesis on the ribosome, the elongation factor Tu (EF-Tu)[3] is the one which in recent years has attracted most attention (for refs see Miller and Weissbach 1977; Bermek 1978; Clark 1980; Parmeggiani and Sander 1980, 1981; Bosch et al. 1983). Because of its remarkable properties and multifunctional activities, EF-Tu can be considered as a model for the investigation of allosteric mechanisms as well as of protein-protein, protein-nucleic acid, and protein nucleotide interactions. The interaction between EF-Tu and aa-tRNA has been object of numerous publications (Gordon 1968; Shorey et al. 1969; Beres and Lucas-Lenard 1973; Ringer and Chladek 1975; Pingoud et al. 1977; Sprinzl et al. 1978; Boutorin et al. 1981; Kao et al. 1983; Wikman et al. 1983). The techniques used were different, but until recently none of them concerned the GTPase activity of the factor. The results of these studies pointed to the importance of the aminoacylation and of the 3' extremity for the interaction with EF-Tu, while leaving many open questions about the participation and role of the individual nucleotidyl residues and sequences of the 3' extremity as well as of the other domains of the aa-tRNA molecule. For this reason, 4 years ago we decided to start the investigation of the interaction between EF-Tu and aa-tRNA by utilizing as a probe the GTP hydrolysis which is one of the fundamental activities of the factor.

1.1 Assay Systems for EF-Tu-Dependent GTPase Activity

For a better understanding of the properties of the assay systems used, we show in Fig. 1 the role of EF-Tu in the elongation cycle. EF-Tu forms a complex with GTP, the prerequisite for the formation of a stable complex with aa-tRNA. The ternary complex EF-Tu·GTP·aa-tRNA thus formed binds to the ribosome programed with mRNA, following codon-anticodon interaction. This reaction is associated with the hydrolysis of the EF-Tu-bound GTP. Since the binding affinity of EF-Tu·GDP for aa-tRNA or the ribosome is physiologically irrelevant, the EF-Tu·GDP complex is released from the

[1] Laboratoire de Biochimie, Ecole Polytechnique, F-91128-Palaiseau Cedex, France
[2] Cattedra di Chimica e Propedeutica Biochimica, 2ª Facolta di Medicina e Chirurgia, Universita di Napoli, I-80131 Naples, Italy
[3] See list of Abbreviations

Mechanisms of Protein Synthesis
ed. by Bermek
© Springer-Verlag Berlin Heidelberg 1984

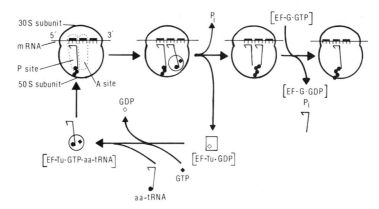

Fig. 1. Schematic representation of the EF-Tu functions in the elongation cycle of polypeptide synthesis. The ribosome·mRNA complex is shown to carry initially a tripeptidyl-tRNA in the P site and finally a tetrapeptidyl-tRNA in the P site. The stimulation by elongation factor Ts of the EF-Tu·GDP/GTP exchange reaction is not illustrated

ribosome·mRNA·aa-tRNA complex. This allows the proper positioning of aa-tRNA into the ribosomal A site and the formation of a new peptide bond between the α-amino group of aa-tRNA and the ester group of peptidyl-tRNA situated in the ribosomal P site. Translocation of the A site-bound peptidyl-tRNA to the P site — a reaction which allows a new cycle of elongation — takes place with the participation of the elongation factor G. The EF-Tu·GDP complex is regenerated to EF-Tu·GTP via an exchange reaction favored by elongation factor Ts. Clearly, the GTP hydrolysis is the crucial reaction which regulates the interaction between EF-Tu and aa-tRNA or the ribosome. The EF-Tu-dependent GTPase also occurs in the presence of unprogramed ribosomes, since the ribosome alone can already activate the catalytic center for GTP hydrolysis situated on EF-Tu. In partial systems containing EF-Tu and ribosomes, but lacking mRNA and/or aa-tRNA, the GTPase activity displays the characteristics of a turnover reaction with linear kinetics. If the EF-Tu concentration is rate limiting, the activity of this system is suitable for evaluating the effects induced by the ribosome and its components (Sander et al. 1980) or by aa-tRNA and its aminoacylated fragments (Parlato et al. 1983) provided that EF-Tu·GTP is efficiently regenerated (Parmeggiani and Sander 1981). To this purpose, we added elongation factor Ts, pyruvate kinase, and phosphoenolpyruvate to our assay system (Parlato et al. 1983). EF-Tu can hydrolyze GTP also in the absence of ribosomes, although in this case the activity is very low and kinetically nonlinear, behaving essentially as a one-round reaction (Fasano et al. 1982).

Another useful system is that utilizing the antibiotic kirromycin. In recent years this drug has been of great help to characterize the properties of EF-Tu (Parmeggiani and Sander 1980). Kirromycin binds in a 1 to 1 ratio to EF-Tu. EF-Tu·kirromycin can interact with GTP and aa-tRNA, and the quaternary complex thus formed binds to the mRNA·ribosome, but after GTP hydrolysis, EF-Tu·GDP remains blocked on the ribosome. This effect is due to the antibiotic action which keeps EF-Tu in a conformation able to interact with aa-tRNA and ribosomes even in the presence of GDP. Moreover, kirromycin activates the EF-Tu site for GTP hydrolysis in the absence of aa-tRNA and ribosomes; both of these two components can strongly increase the antibiotic effect.

In contrast to its behavior in the system containing mRNA-programed ribosomes and aa-tRNA, the GTPase activity of EF-Tu·kirromycin in the absence of codon-anticodon interaction, whether or not ribosomes and/or aa-tRNA are present, displays the characteristics of a linear, turnover reaction (Parmeggiani and Sander 1981) and is thus apt for evaluating the extent of the actions of the different effectors. Since kirromycin strongly stimulates the EF-Tu·GDP/GTP exchange reaction (Fasano et al. 1978), the system with the antibiotic does not require any EF-Tu·GTP regenerating system. The system with kirromycin presents two important features: it allows (a) to investigate the behavior of EF-Tu with respect to GTP hydrolysis in the absence of ribosomes and/or aa-tRNA and (b) to use EF-Tu in a conformation whose constraints regulating its physiological functions are relieved.

2. Methods

Though we used various purified, charged or uncharged tRNAs, and the derived fragments, reaching similar conclusions (Parlato et al. 1981; Parlato et al. 1983; Guesnet et al. 1983; Picone and Parmeggiani 1983), in this communication we limit ourself to the presentation of the results obtained with the $tRNA_1^{Val}$ system except for the experiments described in Fig. 7. Figure 2 shows the various $Val-tRNA_1^{Val}$ 3' fragments obtained by enzymatic digestion with T_1, U_2 and pancreatic A RNases or nuclease S1. They were purified to homogeneity by chromatography and high voltage electrophoresis. In the case of $Val-tRNA_1^{Val}$, the derived fragments were essentially 100% charged, except for $Val-tRNA_1^{Val}$ 3' half molecule (75%-85%). Particular care was taken of checking that the aminoacylated 3' half molecule has been completely devoid of the 5' half molecule (Parlato et al. 1983). Full digestion of the C-C-A 3' end was accomplished with cobra venom phosphodiesterase at pH 9. Pure EF-Tu, elongation factor Ts, tRNA, and tight-couples ribosomes were all from *E. coli*. Hydrolysis of $[\gamma\text{-}^{32}P]GTP$ was followed as liberation of $^{32}P_i$. For details about the purification methods as well as assay procedures, refer to previous publications (Parlato et al. 1981, 1983; Guesnet et al. 1983; Picone and Parmeggiani 1983).

3. Results and Discussion

Our investigation included a broad spectrum of ionic conditions. It is wellknown that cation concentrations, particularly of divalent cations, are critical for optimum activity of the polypeptide synthesis *in vitro* and also for single reactions of this process (Spirin 1973; Chinali and Parmeggiani 1980). In previous papers, we reported in detail about their influence on the EF-Tu-dependent GTPase, with and without kirromycin (Parmeggiani and Sander 1981; Sander et al. 1979; Ivell et al. 1981).

3.1 Assay System in the Presence of Kirromycin

We have first examined the effects on EF-Tu·kirromycin of $Val-tRNA_1^{Val}$ and of short aminoacylated fragments (herein represented by C-C-A-Val) as a function of $MgCl_2$ and

Fig. 2. Val-tRNA$_1^{Val}$ and the derived fragments. The *arrows* indicate the 5' extremity of the different fragments. The Val-tRNA$_1^{Val}$ 3' half molecule consists of a heterogeneous population of aminoacylated fragments containing 38-45 residues as indicated. (From Guesnet J et al. (1983) Eur J Biochem 133: 499-507)

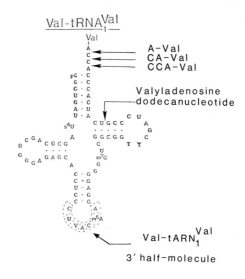

Fig. 3 A-C. Dependence on ammonium ions of the stimulatory effects of Val-tRNA$_1^{Val}$ and C-C-A-Val on the EF-Tu·kirromycin GTPase in the absence of ribosomes at various MgCl$_2$ concentrations. Without (•) or with C-C-A-Val (▲) or Val-tRNA$_1^{Val}$ (■). Blanks of nonenzymatically hydrolyzed [γ-^{32}P]GTP were subtracted. (From Guesnet J et al. (1983) Eur J Biochem 133: 499-507)

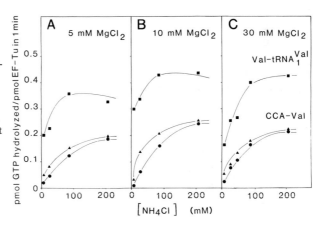

NH$_4$Cl concentration (Fig. 3). The activity of the control EF-Tu·kirromycin was strongly enhanced by increasing NH$_4$Cl concentration. Val-tRNA$_1^{Val}$ greatly stimulated this effect, particularly at low NH$_4$Cl concentration, whereas C-C-A-Val did only a little. This picture did not significantly change at all the three MgCl$_2$ concentrations tested. Presence of ribosomes (not shown, see Parlato et al. 1983) causes (a) a higher basic activity of EF-Tu·kirromycin (four to five times) and (b) a greater stimulatory effect by C-C-A-Val (about half the effect of the intact aa-tRNA molecule). The general picture is the same at 10 and 20 mM MgCl$_2$, whereas at 5 mM MgCl$_2$ the effects of Val-tRNA$_1^{Val}$ and C-C-A-Val are similar and smaller than at 10 mM.

If we study the action of the various aminoacylated oligonucleotides, it can be seen that (a) valyl adenosine is already able to induce a slight stimulation of the EF-Tu·kirromycin GTPase (see Fig. 4A, where the action of the aa-tRNA fragments in the ab-

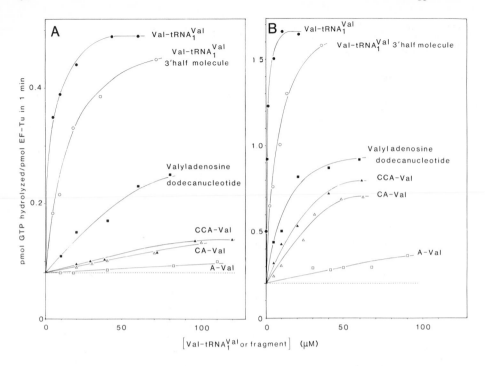

Fig. 4 A-B. Rate of GTP hydrolysis by EF-Tu·kirromycin, without (**A**) or with ribosomes (**B**) as a function of the concentration of Val-tRNA$_1^{Val}$ and its derived aminoacyl-oligonucleotides of different lengths. With Val-tRNA$_1^{Val}$ (●), Val-tRNA$_1^{Val}$ 3' half molecule (○), valyladenosine dodecanucleotide (■), C-C-A-Val (▲), C-A-Val (△), and A-Val (□); no additions (---). (From Guesnet J et al. (1983) Eur J Biochem 133: 499-507)

sence of ribosomes is illustrated as a function of their concentrations and at near-optimum cation concentration); (b) the stimulation is enhanced with increasing the length of the fragment; and (c) in the 3' strand of the acceptor stem there are critical nucleotidyl residues or sequences. The presence of the penultimate cytidylyl residue makes C-A-Val much more active than A-Val and almost as active as C-C-A-Val; the valyladenosine dodecanucleotide is much more effective than C-C-A-Val, whereas the Val-tRNA$_1^{Val}$ 3' half molecule shows the same maximum effect as the whole molecule, albeit at a higher concentration. In the presence of ribosomes (Fig. 4B), the relative stimulations are retained except for the valyladenosine dodecanucleotide, whose stimulation vs. that of C-C-A-Val is much smaller than in the system without ribosomes.

It is important to emphasize that under all conditions tested, in the presence of kirromycin, the effects of the various fragments or of the whole aa-tRNA on the EF-Tu-dependent GTPase were of stimulatory kind.

3.2 Assay System in the Absence of Kirromycin

The effects of the cations in this system display remarkable features. In the experiments in Fig. 5, carried out in the presence of ribosomes, the stimulatory effect of Val-

Fig. 5 A-C. Ammonium dependence of the stimulatory effects of Val-tRNA$_1^{Val}$ ($\bullet$), fragment S$_1$ (=Val-tRNA$_1^{Val}$ 3' half molecule) ($\blacksquare$), and C-C-A-Val ($\blacktriangle$) on the EF-Tu GTPase in the presence of ribosomes at various MgCl$_2$ concentrations, no additions ($\circ$). Blanks of [γ-^{32}P]GTP nonenzymatically hydrolyzed were subtracted. (From Parlato G et al. (1983) J Biol Chem 258: 995-1000)

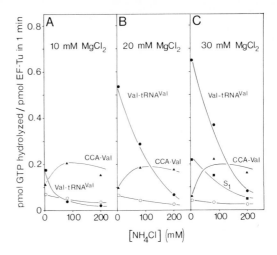

tRNA$_1^{Val}$ increases with increasing MgCl$_2$ and decreases with increasing NH$_4$Cl concentrations to the point that Val-tRNA$_1^{Val}$ inhibits the EF-Tu GTPase at MgCl$_2$ concentrations $\leqslant 10$ mM already in the presence of low quantities of NH$_4$Cl. Interestingly, the Val-tRNA$_1^{Val}$ 3' half molecule displays the same basic characteristics as the whole molecule, though the extent of its effect is three times lower (Fig. 5, shown only for 30 mM MgCl$_2$). By contrast, the stimulatory effect by C-C-A-Val increases with increasing NH$_4$Cl concentrations and is approximately the same at all MgCl$_2$ concentrations.

The differences in the effects of short aminoacylated fragments and of the whole molecule or half molecules become even more evident, when we study the influence of increasing their concentrations at 10, 20, and 30 mM MgCl$_2$ (Fig. 6). At 20 and 30 mM MgCl$_2$, aa-tRNA is stimulatory, whereas at 10 mM it becomes inhibitory. This inhibition is strongly enhanced by further decreasing MgCl$_2$ to 5 mM (see Figs. 8 and 9). By contrast, the short 3' fragments are stimulatory at all MgCl$_2$ concentrations. By examining the individual effects, like in the system with kirromycin, A-Val is slightly stimulatory and C-A-Val induces a much stronger stimulation approaching that of C-C-A-Val; unlike the system with kirromycin, the valyladenosine dodecanucleotide shows a weaker stimulation than C-C-A-Val, while the Val-tRNA$_1^{Val}$ 3' half molecule shows a stimulation at 20 and 30 mM MgCl$_2$ and inhibition at 10 mM, thus supporting the same effect as the whole molecule.

These results prove that besides the fundamental role of the aminoacylated 3' extremity in activating the catalytic center on EF-Tu, there is another region(s) in the aa-tRNA molecule controlling the interaction between the aminoacylated 3' extremity and the EF-Tu center for GTP hydrolysis. Since the 3' half molecule displays the same kind of effects as the whole molecule, we conclude that the TψC loop and stem, the variable loop, and the adjacent regions include part of the regulating region(s). Further analysis of the structures contained in the 3' half molecule, as well as in the whole molecule, presently in progress, may lead to a more detailed characterization of the regulating domain(s) of aa-tRNA.

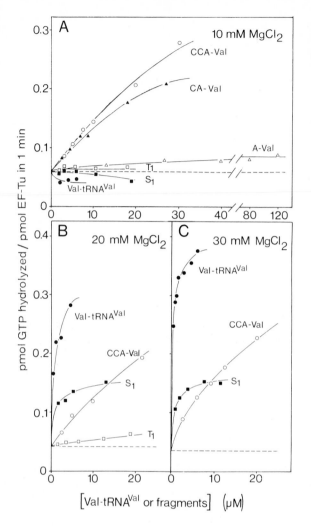

Fig. 6 A-C. Effect on the EF-Tu GTPase of Val-tRNA$_1^{Val}$ and derived aminoacylated fragments of various lengths at different MgCl$_2$ concentrations in the presence of ribosomes. Val-tRNA$_1^{Val}$ (●), fragment S$_1$ (=Val-tRNA$_1^{Val}$ 3' half molecule) (■), fragment T$_1$ (= valyladenosine dodecanucleotide) (□), C-C-A-Val (○), C-A-Val (▲), A-Val (△). The *dashed line* corresponds to the [γ-^{32}P]GTP hydrolyzed by EF-Tu and ribosomes. Blanks of [γ-^{32}P]GTP nonenzymatically hydrolyzed were subtracted. (From Parlato G et al. (1983) J Biol Chem 258: 995-1000)

In the system without kirromycin we have emphasized the results obtained in the presence of ribosomes, since in their absence the EF-Tu-dependent GTPase is very low and nonlinear (see Introduction and Fasano et al. 1982). However, also in the absence of ribosomes, aa-tRNA inhibits the EF-Tu GTPase at low MgCl$_2$, while short aminoacylated fragments do stimulate it (Fig. 7 and Parlato et al. 1983). These results together with those obtained in the system with kirromycin indicate that the described effects depend upon a direct interaction between EF-Tu and aa-tRNA and that they are not mediated via the ribosome which strongly activates the EF-Tu-dependent GTPase reaction.

The observation that MgCl$_2$ regulates the action of aa-tRNA on the EF-Tu GTPase may have physiological importance, since the complete system (containing EF-G and mRNA in addition to EF-Tu, aa-tRNA, ribosomes, and elongation factor Ts) shows a MgCl$_2$ optimum for poly(U)-directed poly(Phe) synthesis at concentrations lower than 10 mM (Chinali and Parmeggiani 1980), that is, in conditions which inhibit the GTPase

Fig. 7. Effect of Lys-tRNA^{Lys} (●) and C-C-A-Lys (▲) on the ammonium-induced one-round GTPase of EF-Tu·[γ-^{32}P]GTP at 10 mM MgCl$_2$ in the absence of ribosomes. EF-Tu alone (□). (From Parlato et al. (1983) J Biol Chem 258: 995-1000)

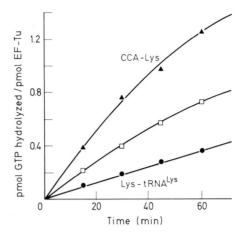

activity of EF-Tu in the ternary complex with or without ribosomes. This would avoid a waste of energy as well as a decrease of the natural substrate for polypeptide synthesis, the ternary complex. Such a situation points to the importance of mRNA-programed ribosomes in inducing the specific orientation of the ternary complex leading to a GTP hydrolysis which is productive for polypeptide synthesis.

The differences observed between the system in the absence of kirromycin and that utilizing the antibiotic can be interpreted as a consequence of the kirromycin action which activated the catalytic site of EF-Tu while influencing the allosteric mechanisms controlling the physiological functions of the factor. With EF-Tu·kirromycin the regulatory domain(s) of aa-tRNA appears to have lost the ability to intervene efficiently and the stimulatory effect on the EF-Tu GTPase remains evident in all conditions.

3.3 Effect of tRNA, Truncated tRNA, and Conclusive Remarks

To get some insight into the mechanisms whereby the regulatory region(s) of aa-tRNA exerts its influence on the aminoacylated 3' extremity, we have extended our studies to the effects induced by tRNAs deprived partially or totally of the 3' C-C-A sequence. We thought that also in this condition one should be able to see some effect of the regulatory region(s), including aa-tRNA domains other than the 3' extremity. We carried out these experiments at low MgCl$_2$ concentration (5 mM), where the regulatory effects are characterized by an inhibition of the EF-Tu GTPase. Figure 8 shows that indeed also uncharged tRNA or truncated tRNA exert an inhibitory effect on the EF-Tu GTPase.

Another series of experiments were performed to examine whether the influence of the regulatory region(s) of tRNA on the GTPase center was acting via a regulatory site(s) of EF-Tu or was reflecting a conformational change induced by MgCl$_2$, modifying the orientation of the aminoacylated 3' extremity with respect to the catalytic center of EF-Tu. To clarify this point, we tested whether the stimulation by C-C-A-Val of the EF-Tu GTPase at low MgCl$_2$ (5 mM) might have been affected by the addition of trun-

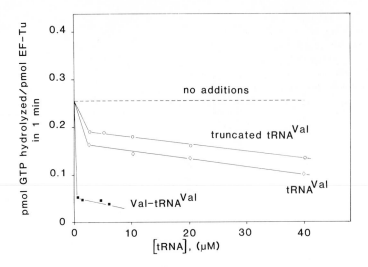

Fig. 8. Effect of Val-tRNA$_1^{Val}$ (■), tRNA$_1^{Val}$ (◇), and tRNA$_1^{Val}$ minus C-C-A (○) on the EF-Tu-dependent GTPase in the presence of ribosomes. The *dashed line* indicates the level of [γ-^{32}P]GTP hydrolyzed by EF-Tu and ribosomes. Blanks of [γ-^{32}P]GTP nonenzymatically hydrolyzed (0.09 pmol GTP/pmol EF-Tu in 1 min) were subtracted. (Modified from Picone, and Parmeggiani (1983) Biochemistry 22, 4400-4405)

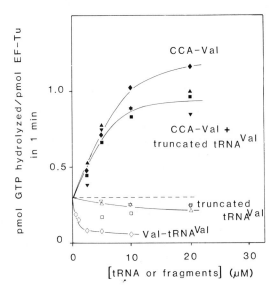

Fig. 9. Influence of different modified tRNA$_1^{Val}$ on the EF-Tu GTPase stimulated by C-C-A-Val in the presence of ribosomes. tRNA$_1^{Val}$ minus pCpCpA-Val (△), minus pA (▽) or minus A (□) at the indicated concentrations. C-C-A-Val at the indicated concentrations, alone (◆) or plus two fold excess of tRNA$_1^{Val}$ minus pCpCpA (▲), minus pA (▼) or minus A (■). The *dashed line* indicates the level of the [γ-^{32}P]GTP hydrolyzed by EF-Tu and ribosomes. Blanks of [γ-^{32}P]GTP nonenzymatically hydrolyzed were subtracted. (Modified from Picone and Parmeggiani 1983) Biochemistry 22, 4400-4405)

cated tRNAs. As shown in Fig. 9, truncated tRNAs can influence the effect of C-C-A-Val, causing a moderate, but significant decrease of its stimulation. Since in this case C-C-A-Val is no longer connected to the remaining tRNA molecule by covalent bonds, such an effect shows that domains of aa-tRNA other than the 3' extremity can interact with EF-Tu and influence via the factor the interaction between 3' extremity and the catalytic center. On the other hand, the observation that the effect of truncated tRNAs came to

Fig. 10. The mechanism regulating the EF-Tu-aa-tRNA interaction with respect to the GTPase activity. The activity of the center for GTP hydrolysis of EF-Tu (GTPase) is enhanced by a ribosomal site(s) (*1*) and by the aminoacylated 3' extremity of aa-tRNA (*2*). A regulatory region(s) of aa-tRNA (*3*) including the TψC loop, TψC stem, extra loop, and vicinal regions, and also other sequences located in the aa-tRNA 5' half molecule influences the GTPase center via a regulatory EF-Tu site(s) (*4*). On its turn, a regulatory EF-Tu site(s) (*4*) affects the conformation of aa-tRNA such to control the productive interaction between the aminoacyl-adenosine of the 3' extremity and the catalytic center of EF-Tu

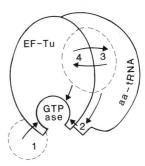

a plateau by increasing their concentrations and that no aa-tRNA-like effect (i.e., an inhibition) could be achieved, strongly suggests that upon binding of aa-tRNA to EF-Tu, the factor itself induces a conformational change of aa-tRNA influencing the orientation of the aminoacylated 3' extremity. This mechanism implies, therefore, a reciprocal adaptation of aa-tRNA and EF-Tu aimed at regulating the degree of productivity of the interaction between the 3' extremity of aa-tRNA and the catalytic center of EF-Tu. Figure 10 is a schematic representation of this model of regulation. It has no claim to represent a true topological situation of the EF-Tu-aa-tRNA interaction with respect to the GTPase center, but only wants to give an idea of the complex mechanism controlling the functional expression of this interaction.

Abbreviations

EF-Tu, elongation factor Tu; GTPase, guanosine 5'-triphosphatase (EC3.6.1.—); aa-tRNA, aminoacyl-tRNA; tRNA$_1^{Val}$, purified isoacceptor 1 of valine-accepting tRNA; Val-tRNA$_1^{Val}$, purified specific tRNA charged with Val; A-Val, 2(3')-*O*-L-valyladenosine; C-A-Val, cytidylyl(3'-5')-2'(3')-*O*-L-valyladenosine; C-C-A-Val, 3' terminal trinucleoside biphosphate charged with Val.

Acknowledgements. This work was supported by NATO Research Grant 1706.

References

Beres L and Lucas-Lenard J (1973) Studies on the fluorescence of the Y base of yeast phenylalanine transfer ribonucleic acid. Effect of pH, aminoacylation, and interaction with elongation factor Tu. Biochemistry 12: 3998-4005

Bermek E (1978) Mechanisms in polypeptide chain elongation on ribosomes. Prog Nucleic Acid Res Mol Biol 21: 63-100

Bosch L, Kraal B, van de Meide PH, Duisterwinkel FJ and van Noort JM (1983) The elongation factor EF-Tu and its two encoding genes. Prog Nucleic Acid Res Mol Biol 30: 91-126

Boutorin AS, Clark BFC, Ebel JP, Kruse TA, Petersen HU, Remy JP and Vasilenko S (1981) A study of the interaction of *Escherichia coli* elongation factor Tu with aminoacyl-tRNAs by partial digestion with cobra venom ribonuclease. J Mol Biol 152: 593-608

Chinali G and Parmeggiani A (1980) The coupling with polypeptide synthesis of the GTPase activity dependent on elongation factor G. J Biol Chem 253: 7455-7459

Clark BFC (1980) The elongation step of protein biosynthesis. Trans Biochem Sci 5: 207-210

Fasano O, De Vendittis E and Parmeggiani A (1982) Hydrolysis of GTP by elongation factor Tu can be induced by monovalent cations in the absence of other effectors. J Biol Chem 237: 3145-3150

Fasano O, Bruns W, Crechet JB, Sander G and Parmeggiani A (1978) Modification of elongation factor Tu-guanine nucleotide interaction by kirromycin. Eur J Biochem 89: 557-565

Gordon J (1968) A stepwise reaction yielding a complex between a supernatant fraction from *E. coli,* guanosine 5' triphosphate and aminoacyl-sRNA. Proc Natl Acad Sci USA 59: 179-183

Guesnet J, Parlato G and Parmeggiani A (1983) Interaction between the different domains of aminoacyl-tRNA and the elongation factor Tu·kirromycin complex. Eur J Biochem 13: 499-507

Ivell R, Sander G and Parmeggiani A (1981) Modulation by monovalent cations of the guanosine 5' triphosphatase activity dependent on elongation factor Tu. Biochemistry 20: 6852-6859

Kao TH, Miller DL, Abo M and Ofengand J (1983) Formation and properties of a covalent complex between elongation factor Tu and Phe-tRNA bearing a photoaffinity labeling probe on its 3-(3-amino-3-carboxypropyl)uridine residue. J Mol Bio. 166: 383-405

Miller D, Weissbach H (1977) Aminoacyl-tRNA transfer factors. In: Weissbach H and Pestka S (eds) Molecular mechanisms of protein synthesis. Academic, New York, p 323-373

Parlato G, Guesnet J, Crechet JB and Parmeggiani A (1981) The GTPase activity of elongation factor Tu and the 3' terminal end of aminoacyl-tRNA. FEBS Lett 125: 257-260

Parlato G, Pizzano R, Picone D, Guesnet J, Fasano O and Parmeggiani A (1983) Different regions of aminoacyl-tRNA regulate the function of elongation factor Tu. J Biol Chem 258: 995-1000

Parmeggiani A and Sander G (1980) Properties and action of kirromycin (Mocimycin) and related antibiotic. In: Sammes PG (ed) Topics in antibiotic chemistry, vol 5. Wiley, Chichester, England, pp 159-221

Parmeggiani A and Sander G (1981) Properties and regulation of the GTPase activities of elongation factor Tu and G, and of initiation factor 2. Mol Cell Biochem 31: 129-158

Picone D and Parmeggiani A (1983) Transfer ribonucleic acid deprived of the C-C-A 3' extremity can interact with elongation factor Tu. Biochemistry 22: 4400-4405

Pingoud A, Urbanke C, Krauss G, Peters F and Maas G (1977) Ternary complex formation between elongation factor Tu, GTP and aminoacyl-tRNA: an equilibrium study. Eur J Biochem 78: 403-409

Ringer D and Chládek S (1975) Interaction of elongation factor Tu with 2'(3')-*O*-aminoacyloligonucleotides derived from the 3' terminus of aminoacyl-tRNA. Proc Natl Acad Sci USA 72: 2950-2974

Sander G, Okonek M, Crechet JB, Ivell R, Bocchini V and Parmeggiani A (1979) Hydrolysis of GTP by the elongation factor Tu·kirromycin complex. FEBS Lett 98: 111-114

Shorey RL, Ravel JM, Garner CW and Shive W (1969) Formation and properties of the aminoacyl transfer ribonucleic acid-guanosine triphosphate-protein complex. J Biol Chem 244: 4555-4564

Spirin A (1973) On the structural basis of ribosome functioning. In: Zelinka J, Balan J (eds) Ribosomes and RNA metabolism Vol 1. Publishing House of the Slovak Academy of Sciences, Bratislava, pp 33-68

Sprinzl M, Siboska GE and Petersen JA (1978) Properties of tRNA[Phe] from yeast carrying a spin label on the 3' terminal. Interaction with yeast phenylalanyl-tRNA synthetase and elongation factor Tu from *Escherichia coli.* Nucleic Acids Res 5: 861-877

Wikman FP, Siboska GE, Petersen HU and Clark BFC (1983) The site of interaction of aminoacyl-tRNA with elongation factor Tu. EMBO J 9: 1095-1100

Recognition of Aminoacyl-tRNA by Bacterial Elongation Factor Tu

M. Sprinzl and H. Faulhammer[1]

1. Introduction

The formation of a ternary complex between aminoacyl-tRNA, elongation factor Tu (EF-Tu), and GTP is an obligatory step during the peptide chain elongation. Despite the available information on tertiary structure of tRNA, little is known about the molecular details of this interaction.

It is assumed that the distribution of codons in mRNA together with the distribution of corresponding anticodons of tRNA isoacceptors have an influence on the rate and error frequency of translation (Konigsberg and Godson 1983). If this is the case, then the participation of a certain anticodon in the translation process will depend on the cellular concentration of the particular ternary complex. That means, that not only the concentration of the tRNA isoacceptors should be considered, but also their ability to form ternary complexes with EF-Tu·GTP. In order to investigate the hypothesis of the translational control by codon usage, a precise method for the quantification of ternary complexes must be available. We approached this problem by a systematic study of the interaction of aminoacyl-tRNAs with the EF-Tu. The central question in these investigations was not a topographic study of mutual contact sites between the nucleic acid and the protein (Boutorin et al. 1981; Wikman et al. 1982), but rather the question on the efficiency of interaction between aminoacyl-tRNA and EF-Tu, its modulation by guanosine nucleotides, aminoacyl-tRNA structure, and posttranscriptional modification.

The following methods were used in these investigations: specific chemical modification of aminoacyl-tRNAs and guanosine nucleotides, introduction of spectroscopic labels into tRNA molecules followed by physicochemical measurements, and affinity chromatography of the aminoacyl-tRNAs on immobilized EF-Tu·GTP.

2. Interaction of Aminoacyl-tRNA With EF-Tu·GTP

2.1 Role of the Aminoacyl Residue

The aminoacyl residue of the aminoacyl-tRNA is important for EF-Tu binding, although a very weak interaction has also been observed with deacylated tRNAs (Miller

[1] Universität Bayreuth, Lehrstuhl für Biochemie, Postfach 3008, D-8580 Bayreuth.

Mechanisms of Protein Synthesis
ed. by Bermek
© Springer-Verlag Berlin Heidelberg 1984

and Weissbach 1977; Jonak et al. 1980; Picone and Parmeggiani 1983). The α-amino group of the aminoacyl residue is apparently directly involved in binding to the protein, since in ternary complex it is protected against chemical acylation (Sedlacek et al. 1976).

The simple replacement of the α-amino group of Phe-tRNA^{Phe} by a hydroxyl group after nitrous acid treatment leads to a considerably lower affinity of Phelac-tRNA^{Phe} to EF-Tu·GTP. The apparent dissociation constant of the Phelac-tRNA^{Phe} (Fig. 1a) is higher by a factor of 300 than the K_D^{app} of the native Phe-tRNA^{Phe} (Fig. 1 b). The complete elimination of the α-amino group leads to the formation of cinnamyl-tRNA^{Phe} (Fig. 1 c), which has about the same affinity ($K_D^{app} = 3 \times 10^{-5}$ M) to EF-Tu·GTP as the Phelac-tRNA^{Phe} (Derwenskus and Sprinzl 1983). Thus, the presence of the α-hydroxyl group in Phelac-tRNA^{Phe} does not have any stabilizing influence on the formation of the ternary complex. All modifications on the α-amino group, including N-acylation (Ravel et al. 1967), replacement by a hydroxyl group, or complete elimination, lead to an identical reduction of the affinity of the modified aminoacyl-tRNAs to EF-Tu·GTP. It is, therefore, likely that the α-amino group has an important function upon complex formation with EF-Tu which can not be fulfilled by an α-hydroxyl group.

The dissociation constants of native aminoacyl-tRNAs in the interaction with EF-Tu·GTP are not the same for all tRNAs. Differences between particular isoacceptors which are greater than one order of magnitude were measured (Pingoud and Urbanke 1980). The observed differences in the dissociation constants may be due either to the structure of the tRNA or to the nature of the amino acid side chain. The fact that the side chain of the aminoacyl residue indeed influences the strength of the interaction of aminoacyl-tRNA with EF-Tu·GTP was demonstrated by measurements of the interaction of Lys-tRNA^{Lys} and misaminoacylated Phe-tRNA^{Lys} with EF-Tu·GTP which were differing only in the amino acid side chain. It was observed that K_D^{app} is increased by a factor of 20 for Lys-tRNA^{Lys} as compared to Phe-tRNA^{Lys}. Thus, in this case, the misaminoacylated Phe-tRNA^{Lys} binds more efficiently to EF-Tu·GTP as the Lys-tRNA^{Lys} (Wagner and Sprinzl 1980). There are two possible explanations for this observation. As pointed out by Pingoud and Urbanke (1980) there could be a pocket for the amino acid side chain in the EF-Tu, which binds preferentially the aromatic amino acids. The stereospecificity of the interaction of L-aminoacyl-tRNAs with EF-Tu·GTP is in support of this suggestion, since it was shown that D-aminoacyl-tRNAs bind much more weakly to EF-Tu·GTP than the L-aminoacyl-tRNAs (Calendar and Berg 1966; Pingoud and Urbanke 1980; Yamane et al. 1981; Bhuta et al. 1982). Another possibility which could explain the effect of the side chain of the aminoacyl residue on

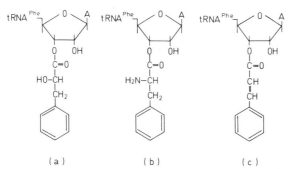

Fig. 1 a-c. Structure of the 3' end of a Phelac-tRNA^{Phe}; b Phe-tRNA^{Phe}; and c cinnamyl-tRNA^{Phe}. tRNA species a and c were obtained by treatment of Phe-tRNA^{Phe} from *E. coli* by nitrous acid (Derwenskus and Sprinzl 1983)

ternary complex formation is the requirement for a protonated α-amino group for this interaction. The protonation of the α-amino group of the aminoacyl-tRNA will obviously depend on the nature of the amino acid side chain since the pK values of the α-amino group of aminoacylesters are dependent on the structure of the amino acid (Hay and Porter 1967).

The hydroxyl function neighbouring the aminoacyl group evidently plays some role 3' position of the terminal adenosine (Taiji et al. 1983). Therefore, at least two possible substrates can be postulated for the interaction of aminoacyl-tRNA with EF-Tu·GTP. The question which of the two isomers is involved in ternary complex formation was investigated in several laboratories using nonisomerizable aminoacyl-tRNA analogues, shown in Fig. 2. Sprinzl et al. (1977) and Hecht et al. (1977) demonstrated that both isomeric "deoxy" aminoacyl-tRNA species, Fig. 2 d and e can form ternary complexes with EF-Tu·GTP, although with considerably lower efficiency as the corresponding native aminoacyl-tRNA. There is no clear preference, however, for one of the positional isomers.

The hydroxyl function neighbouring the aminoacyl group evidently plays some role in binding to EF-Tu. In analogues, shown in Fig. 2 f and g the aminoacylester linkage is replaced by the aminoacylamido bond. No interaction of these species with EF-Tu can be detected. It is possible that conformational differences caused by the presence of either amido or ester bond can account for these observations. Furthermore, it could be postulated that the aminoacyl-tRNA during its interaction with EF-Tu·GTP adopts an intermediary orthoacid structure as depicted in Fig. 2 b. The partially unsaturated character of the amido linkage leading to lower susceptibility of the carbonyl group to nucleophilic reaction will make such an orthoacid structure in the case of "deoxyamino" aminoacyl-tRNAs very unlikely. A weaker binding of the isomers, shown in Fig. 2 d and e to EF-Tu·GTP seems to support this hypothesis.

2.2 Role of the tRNA Structure

The involvement of the CCA terminus of aminoacyl-tRNA in the interaction with EF-Tu·GTP is a crucial one, although other parts of tRNA are probably playing important roles as well. There are several indications that other regions of aminoacyl-tRNA interact directly with EF-Tu·GTP (Boutorin et al. 1981; Bhuta et al. 1982; Parlato et al. 1983; Adkins et al. 1983; Clark et al. 1983; Picone and Parmeggiani 1983; Parmeggiani, this book). One such region is, e.g., the double stranded acceptor stem adjacent to the CCA sequence (Schulman et al. 1974). It was also suggested that the 5' phosphate group of the aminoacyl-tRNA may form an intramolecular salt bridge with the α-amino group of the aminoacyl residue and that such a feature may be important for the recognition of the aminoacyl-tRNA by EF-Tu·GTP (Schulman et al. 1974). However, this hypothesis was found to be incorrect, since there is no difference between the binding of native aminoacyl-tRNA and 5' dephosphorylated aminoacyl-tRNA to EF-Tu·GTP (Sprinzl and Graeser 1980). The possible participation of the variable loop of aminoacyl-tRNA in ternary complex formation was studied in our laboratory. tRNA^{Phe} from E. coli (Fig. 3) contains in the variable loop a modified nucleoside, 3-N(3-amino-3-carboxypropyl)uridine (base X-47), to which a fluorescent group can be attached by a selective reaction with

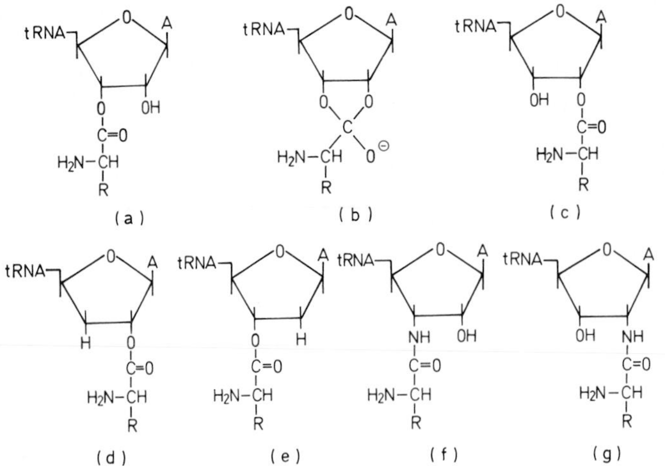

Fig. 2 a-g. Structure of the 3' end of aminoacyl-tRNA; native aminoacyl-tRNA with the aminoacyl residue attached to the 3' position of the terminal adenosine a, the intermediary orthoacid structure of the aminoacyl-tRNA b, and the 2' aminoacylated isomer c. The nonisomerizable analogues used in this study are the aminoacyl-tRNA(3'-deoxy)A d, aminoacyl-tRNA(2'-deoxy)A e, aminoacyl-tRNA(3'deoxy-3'amino)A f, and the aminoacyl-tRNA (2'-deoxy-2'-amino)A g

fluorescamine (Sprinzl and Faulhammer 1978). Such a modified tRNA^{Phe} (XF) can be aminoacylated and as such interacts with EF-Tu·GTP as efficiently as the native Phe-tRNA^{Phe}. In the case of phenylalanyl-tRNA synthetase the binding of tRNA^{Phe} (XF) to the enzyme leads to a pronounced change in the fluorescence amplitude. No such effect can be observed when Phe-tRNA^{Phe} (XF) binds to EF-Tu·GTP. We, therefore, conclude from this investigation that the variable loop of Phe-tRNA^{Phe} (XF) is probably not involved in the interaction with EF-Tu·GTP. This conclusion was confirmed in recent reports (Pingoud and Urbanke 1980; Boutorin et al. 1981).

Similar results as those observed with Phe-tRNA^{Phe} (XF) were also obtained by measurements of the fluorescence of the modified nucleoside Y (wyosine) in position 37 of tRNA^{Phe} from yeast (Fig. 3). The removal of this nucleoside does not prevent the formation of a Phe-tRNA^{Phe} (-Y) ternary complex with EF-Tu·GTP. If the natural fluorescence of the Y-base is measured, no change was detected after the binding of Phe-tRNA^{Phe} to EF-Tu·GTP (Beres and Lucas-Lenard 1973). Analogous observations were made when an aminoacylated tRNA^{Tyr} from E. coli containing a modified Q-34 base was tested in ternary complex formation (Pingoud et al. 1976).

In order to study the influence of EF-Tu·GTP on the conformation of the anti-codon loop of aminoacyl-tRNA in more detail a stable nitroxide spin label (Fig. 3) was attached to 2-thiocytidine in position 32 of the anticodon loop of tRNA^{Arg} from E. coli (Kruse et al. 1978). The spin-labelled Arg-tRNA^{Arg} (SL) formed ternary complexes with EF-Tu·GTP. Electron spin resonance (ESR) spectra of modified tRNA^{Arg} (SL), Arg-tRNA^{Arg} (SL), and Arg-tRNA^{Arg} (SL). EF-Tu·GTP ternary complexes were then measured in order to identify possible conformational changes in the tRNAs depending on different functional states. In Table 1 the rotational correlation times of the spin

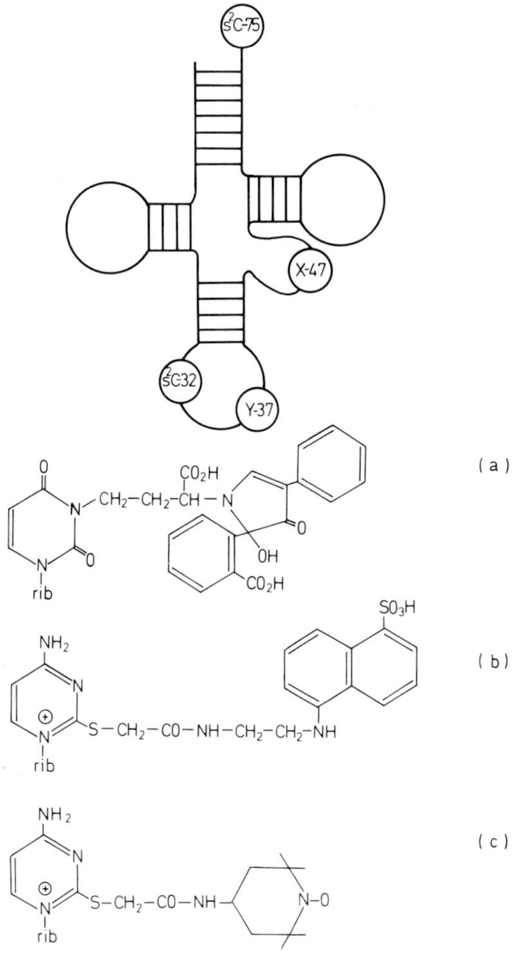

Fig. 3 a-c. Cloverleaf structure of tRNA showing modified base residues. The fluorescamine label **a** was introduced into position 47 of tRNA^{Phe} from *E. coli*. The AEDANS label **b** was introduced either to position 32 of tRNA^{Arg} from *E. coli* or to position 75 of tRNA^{Phe} from yeast. Spin labelling **c** was performed on tRNA^{Arg} from *E. coli* (position 32) or tRNA^{Phe} from yeast (position 75), respectively. The Y-37 is a natural modified nucleoside in tRNA^{Phe} from yeast

labels attached to tRNA^{Arg} are presented. Whereas the aminoacylation does not influence the mobility of the nitroxid spin label attached to position 32, the formation of the Arg-tRNA^{Arg} (SL). EF-Tu·GTP complex leads to a slight immobilization of the spin label. This study reveals that although there is no direct interaction of the EF-Tu with the anticodon loop of the tRNA, the formation of the aminoacyl-tRNA·EF-Tu·GTP complex does affect the conformation of the anticodon loop. A similar approach, using spin-labelled tRNAs was applied to investigate the participation of the CCA-end of tRNA in the interaction with EF-Tu·GTP. Replacement of the invariant penultimate cytidine residue by cytidine analogues, e.g. 2-thiocytidine, 5-azacytidine, or 5-iodocytidine, respectively, does not impair the ternary complex formation. Schulman et al. (1974) reported that an aminoacyl-tRNA in which the cytidine-75 is replaced by uridine can bind only weakly to EF-Tu·GTP. Thus, modification of carbon 4 in the pyrimidine ring of cytidine-75 is not tolerated by EF-Tu. On the other hand, the alkylation of s^{2}C-75 in Phe-tRNA^{Phe} with a bulky substituent, produces aminoacyl-tRNA species fully

Table 1. Rotational correlation times τ of a spin label in tRNA^{Arg} from *E. coli* and tRNA^{Phe} from yeast, respectively[a]

Spin labelled nucleotide:s^2C-32		Spin-labelled nucleotide:s^2C-75	
Sample	τ [ns]	Sample	τ [ns]
tRNA^{Arg}(SL)	1.87	tRNA^{Phe}(SL)	1.36
Arg-tRNA^{Arg}(SL)	1.91	Phe-tRNA^{Phe}(SL)	0.76
Arg-tRNA^{Arg}(SL)·EF-Tu·GTP	2.71	Phe-tRNA^{Phe}(SL)·EF-Tu·GTP	0.66

[a]The ESR spectra were recorded at 28°C. The structure of the respective spin-labelled tRNA is depicted in Fig. 3 c. The data are taken from Kruse et al. (1978) and Sprinzl et al. (1978)

active in the ternary complex formation (Sprinzl et al. 1978). Such a modified tRNA was obtained by alkylation of yeast tRNA^{Phe}-C-s^2C-A with a spin label reagent and was used for ESR-spectroscopic experiments. The rotational correlation times of the spin-labelled tRNA^{Phe} (SL), Phe-tRNA^{Phe} (SL), and Phe-tRNA^{Phe} (SL)·EF-Tu·GTP ternary complexes were determined (Table 1). The spin label in the aminoacylated Phe-tRNA^{Phe} (SL) has increased mobility as compared to tRNA^{Phe} (SL), indicating that the amino-acylation influences the structure of the single-stranded CCA region. The binding of EF-Tu·GTP surprisingly does not affect the motion of the spin label attached to residue 75. It was, therefore, concluded from these experiments that the s^2C-75 residue is not involved in the interaction with EF-Tu. Probably the EF-Tu does not interact with the aglycon moieties of the CCA-end, but recognizes instead the overall spacial outline of the stacked CCA-terminus, leaving the nucleobases of the CCA-end free for possible interaction with ribosomal A-site in the next step of the peptide chain elongation pro-cess (Küchler, this book).

2.3 Determination of the Binding Constants

The dissociation constants of the interaction between aminoacyl-tRNA and EF-Tu·GTP can be determined by indirect methods, such as the nitrocellulose filter binding assay (Miller and Weissbach 1977), the hydrolysis protection assay (Pingoud and Urbanke 1980), and the RNAse-resistance assay (Tanada et al. 1982). Recently an optical detec-tion method became available following the change of an optical parameter as a result of ternary complex formation. In this work the fluorescence of a fluorescein moiety attached to the s^4U-8 base of Phe-tRNA^{Phe} from *E. coli* was used as an optical signal. The change of the fluorescence amplitude upon ternary complex formation was inter-preted in terms of an indirect interaction of EF-Tu with the fluorescein reporter group, indicating a conformational change in Phe-tRNA^{Phe} near the base s^4U-8 induced by EF-Tu·GTP (Adkins et al. 1983).

In a similar approach we utilized a functionally active yeast Phe-tRNA^{Phe} with a covalently bound naphthylamine sulfonic acid (AEDANS) reporter group at the 3' terminal acceptor end. The fluorescent AEDANS residue was introduced into a modi-fied tRNA^{Phe}-C-s^2C-A by alkylation of 2-thiocytidine in position 75 with a iodo-

Fig. 4. Fluorescence spectrum of free Phe-tRNA^{Phe} (AEDANS)-A, alone (*broken line*), or in a complex with EF-Tu·GTP (*solid line*). The spectrum was measured in aqueous solution, the excitation wavelength was 335 nm

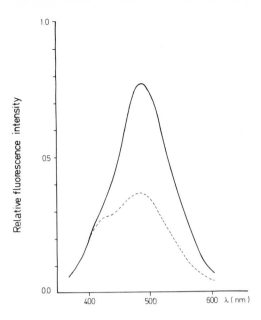

acetamido analogue of AEDANS (Fig. 3). The interaction of Phe-tRNA^{Phe}-C-s^{2}C (AEDANS) A with a twofold molar excess of EF-Tu·GTP from *E. coli* leads to approximately 100% fluorescence enhancement at 485 nm (Fig. 4). From the evaluation of a fluorescence titration of Phe-tRNA^{Phe}-C-s^{2}C(AEDANS)-A with EF-Tu·GTP a dissociation constant of $5\pm2 \times 10^{-8}M$ for the ternary complex formation could be calculated. Since the binding of the modified aminoacyl-tRNA results in such a drastic increase of the fluorescence quantum yield relative to the AEDANS dye in uncomplexed Phe-tRNA^{Phe}-C-s^{2}C(AEDANS)A, it appears reasonable to assume that the 3' terminal acceptor end of aminoacyl-tRNA bound to EF-Tu·GTP is located at a binding site of low polarity. This could be a groove or a pocket with an extensive exclusion of water. Such an environment in the proximity of the 3' terminal adenosine could contribute to the apparent stabilization of the amino acid ester bond of aminoacyl-tRNA complexed by EF-Tu·GTP.

There is no change of fluorescence, however, if EF-Tu·GTP is replaced by EF-Tu·GDP complex. On the other hand, fluorescence polarization measurements indicate that EF-Tu·GDP is able to bind the aminoacylated tRNA^{Phe}-C-s^{2}C(AEDANS)-A, albeit much more weakly than EF-Tu·GTP. It is, thus, assumed that the 3' terminal acceptor end of the modified Phe-tRNA^{Phe} bound to EF-Tu·GTP must be in a completely different local environment as compared to EF-Tu·GDP.

Other nonfluorescent aminoacyl-tRNAs as, e.g., Arg-tRNA^{Arg} from *E. coli*, Tyr-tRNA^{Tyr}, or Phe-tRNA^{Phe} (lacking Y-37 base) from yeast compete with the fluorescent tRNA^{Phe}-C-s^{2}C(AEDANS)-A for the aminoacyl-tRNA binding site on EF-Tu·GTP. The resulting fluorescence decrease can then be used for the determination of the dissociation constant of any aminoacyl-tRNA via competition assay.

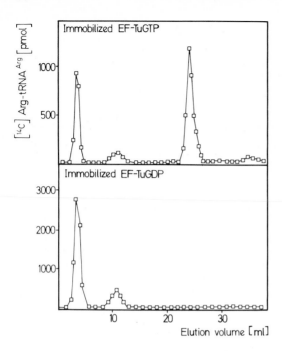

Fig. 5. Affinity chromatography of Arg-tRNA[Arg] from *E. coli* on immobilized EF-Tu. Unfractionated tRNA from *E. coli* was aminoacylated with [^{14}C] arginine and passed through a column of immobilized EF-Tu·GTP (*above*) or EF-Tu·GDP (*below*). Three types of buffer were used for the elution; up to 5 ml elution volume, a buffer of low ionic strength, followed by a buffer of intermediate ionic strength (up to 20 ml elution volume), and then a buffer of high ionic strength. The small peak at 11 ml elution volume in the case of the immobilized EF-Tu·GDP column and the peak at 25 ml in the case of immobilized EF-Tu·GTP column, contain the Arg-tRNA[Arg] species. tRNA[bulk] appears in the break through fraction of the column (3 ml)

3. Affinity Chromatography of Aminoacyl-tRNAs on Immobilized EF-Tu·GTP

Binding of aminoacyl-tRNAs to EF-Tu·GTP can be easily monitored by affinity chromatography using an immobilized EF-Tu·GTP (Fischer et al. 1982). The elongation factor Tu can be covalently bound to cyanogen bromide-activated sepharose. Although a considerable amount of EF-Tu is deactivated during the coupling procedure, the remaining active part of the immobilized protein retains all properties of the soluble EF-Tu. It binds GDP and GTP and does not interact with other nucleotides. Immobilized EF-Tu·GDP can be converted to EF-Tu·GTP, which binds aminoacyl-tRNAs (Fig. 5). Several features of the bacterial elongation factor Tu could be investigated by affinity chromatography:

1. EF-Tu·GDP although much less efficiently than EF-Tu·GTP, interacts with aminoacyl-tRNAs (Fig. 5).

2. The efficiency of the interaction of aminoacyl-tRNA with EF-Tu·GTP varies with the particular isoacceptors. For instance, if a group of Arg-tRNA^{Arg} isoacceptors from *E. coli* are bound to a column of immobilized EF-Tu·GTP and then eluted stepwise with buffers of increasing ionic strength, the distribution of Arg-tRNA^{Arg} species in the eluate is different from one step to the other. Therefore, single Arg-tRNA^{Arg} isoacceptors bind with different affinity.

3. The efficiency of the interaction of aminoacyl-tRNAs with EF-Tu·GTP is different for the particular group of tRNA isoacceptors. For example, the binding of Phe-tRNA^{Phe} species from yeast to immobilized EF-Tu·GTP from *Thermus thermophilus* is much less efficient than the binding of Tyr-tRNA^{Tyr} isoacceptors from the same source.

4. Bacterial EF-Tu can bind yeast mitochondrial tRNAs, yeast cytoplasmic tRNAs, and mammalian cytoplasmic tRNAs.

The immobilized EF-Tu·GTP can be used for the isolation of any group of aminoacyl-tRNA isoacceptors from unfractionated tRNA. For this purpose the unfractionated tRNA is charged with a crude aminoacyl-tRNA synthetase in the presence of one radioactively labelled amino acid. The resulting reaction mixture is passed through a column of immobilized EF-Tu·GTP. Only the aminoacylated tRNAs are retained which can be eluted from the column with a buffer of high ionic strength (Fig. 5) and analyzed by two-dimensional polyacrylamide gel electrophoresis (Fig. 6). With this method the interaction of EF-Tu·GTP with other tRNA-like molecules as for instance plant viral RNAs can be investigated. Moreover the distribution of tRNA isoacceptors in complicated cellular extracts can easily be determined.

4. Modulation of the EF-Tu Structure by Guanosine Nucleotides

Bacterial elongation factor Tu binds GDP and GTP (Kaziro 1978) and shows virtually no affinity for GMP or other nucleotides (Arai et al. 1978).

The EF-Tu·GTP complex recognizes and binds aminoacyl-tRNA with high specificity, but not elongation factor Ts (Miller and Weissbach 1977; Kaziro 1978) and can interact with the ribosomal A-site even in the absence of aminoacylated tRNA (Kawakita et al. 1974). The EF-Tu·GDP complex has no affinity for tRNA and ribosomes, but strongly binds elongation factor Ts (Kaziro 1978). These observations were explained by two different conformations of elongation factor Tu, which apparently were induced by the action of the bound GDP or GTP nucleotide.

Evidence for such conformational differences between EF-Tu·GDP and EF-Tu·GTP has been obtained by several physicochemical methods, such as ESR (Arai et al. 1976; Wilson et al. 1978), NMR (Nakano et al. 1980) intrinsic protein fluorescence (Arai et al. 1977), fluorescence of reporter groups (Crane and Miller 1974; Arai et al. 1975), by chemical methods, such as hydrogen exchange (Printz and Miller 1973; Ohta et al. 1977), and by enzymatic treatment (Douglass and Blumenthal 1979).

The structural requirements of the GDP/GTP binding site of bacterial elongation factor Tu have been the subject of extensive investigations using several natural and synthetic guanosine nucleotide analogues. Modifications at various positions of the

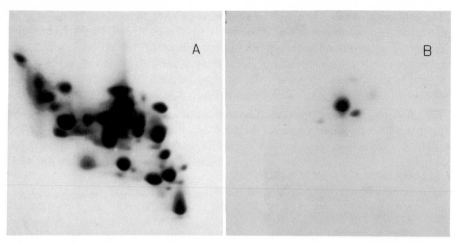

Fig. 6 A, B. Two-dimensional electrophoresis on polyacrylamide gels of *E. coli* tRNA[bulk] originating from **A** break–through fraction of the affinity chromatography described in Fig. 5. The electrophoretic analysis of the tRNA[Arg] isoacceptors isolated by affinity chromatography on an immobilized EF-Tu·GTP column is shown in **B**

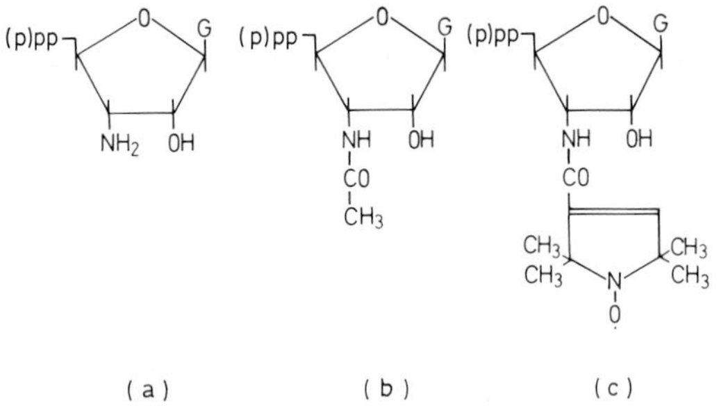

Fig. 7 a-c. Structure of the analogues of guanosine-5'-diphosphate and guanosine-5'-triphosphate, derived from **a** 3'-deoxy-3'-amino-guanosine and **b** 3'-deoxy-3'-acetaminoguanosine. Spin-labelled analogues 3' (SL) GDP and 3' (SL) GTP **c**

purine ring or at the ribose were introduced and the complex formation between the nucleotide analogue and elongation factor Tu was studied (Miller and Weissbach 1977; Wittinghofer et al. 1977; Eccleston 1981).

   Potential fluorescent or spin-labelled analogues which are useful analytical tools for the examination of protein structure and function can be derived from amino-modified guanosine nucleotides. Substitution of the 3' hydroxyl group by a 3' amino group in nucleotides (Fig. 7) was recently achieved chemically (Azhayev et al. 1979) and the resulting 3' (NH$_2$) GDP and 3' (NH$_2$) GTP were analyzed with respect to their substrate properties towards *E. coli* EF-Tu (Table 2). As can be seen the replacement of the 3' hydroxyl group in the ribose moiety by a 3' amino group changes the binding affinity of 3' (NH$_2$) GDP and 3' (NH$_2$) GTP only to a very little extent. The amino group of 3' (NH$_2$) GDP or 3' (NH$_2$) GTP can be conveniently used for the introduction of spin label reporter groups. It can be further shown that the additional acetylation or even acylation of the 3' amino group with the rather bulky 2, 2, 5, 5-tetramethyl-3-pyrrolin-1-oxyl-3-

Table 2. Dissociation constants K_D^{app} of guanosine nucleotides and their analogues in the complex formation with EF-Tu from *E. coli*[a]

Guanosine nucleotides	$K_D^{app}[M]$
GDP	3×10^{-9}
GTP	330×10^{-9}
3' (NH$_2$)GDP	12×10^{-9}
3' (SL) GDP	9×10^{-9}
3' (NH$_2$)GTP	330×10^{-9}
3' (SL) GTP	750×10^{-9}

[a]The formulae are given in Fig. 7

carboxylic acid (Fig. 7) yielding 3' (SL) GDP or 3' (SL) GTP only moderately reduces the affinity of the analogues towards *E. coli* EF-Tu (Table 2). This experimental finding is in support of previous observations that the ribose ring of bound guanosine nucleotides is partly buried, but with the 2', 3'-hydroxyl groups exposed to the solvent (Duister-winkel et al. 1984).

In order tu study the nature of the protein-nucleotide interaction, ESR measurements were performed. The ESR-spectrum of uncomplexed 3' (SL) GDP in aqueous solution is given in Fig. 8. The typical three line pattern characteristic for a fully mobile nitroxide spin label (rotational correlation time τ about 0.1 ns) is shown with a center field peak of g = 2.008. Upon addition of *T. thermophilus* EF-Tu to 3' (SL) GDP or 3' (SL) GTP an intensive broadening of the hyperfine lines in both ESR spectra is observed reflecting an anisotropic motion of the label in EF-Tu·3' (SL) GDP and EF-Tu· 3' (SL) GTP. The occupation of the binding site of EF-Tu by the spin-labelled guanosine nucleotides results in a strong motional broadening of the hyperfine lines especially that of the high field resonance. The shape of the lines corresponds to the ESR-spectra of moderately immobilized nitroxide spin labels (rotational correlation time τ about 10 ns) with the motion of the bound label fairly correlating with the overall motion of the macromolecule.

The binding of Tyr-tRNA^{Tyr} from yeast to the binary EF-Tu·3' (SL) GTP complex resulting in ternary Tyr-tRNA^{Tyr}·EF-Tu·3' (SL)GTP complex formation can also be followed by ESR spectroscopy. The ternary complex is likewise characterized by a significant line broadening in the high field region of the ESR spectrum (Fig. 8). In addition, an inversion of the amplitude of the high field resonance can be detected which is not present in the investigated binary complexes.

These observations show that the environment of the ribose moiety of the guanosine nucleotides in EF-Tu·3' (SL)GDP, in EF-Tu·3'(SL)GTP and in aminoacyl-tRNA·EF-Tu· 3'(SL)GTP ternary complex is different. Such effects could be caused by changes of the polarity in the direct vicinity of the spin label or by restrictions in the molecular motion of the ESR reporter groups.

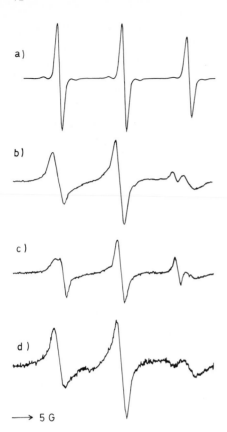

Fig. 8 a-d. ESR-spectra of **a** 3' (SL) GDP; **b** EF-Tu·3' (SL) GDP; **c** EF-Tu·3' (SL) GTP; and **d** Tyr-tRNA^{Tyr}·EF-Tu·3' (SL) GTP. The spectra (X-band spectra, 9.7 GHz) were recorded at room temperature

a)

b)

c)

d)

⟶ 5 G

5. Conclusions

Elongation factor Tu changes its conformation upon binding guanosine nucleotides. This structural change can be monitored by measurement of ESR spectra of the corresponding complexes using spin-labelled guanosine nucleotides. The binding of aminoacyl-tRNA to EF-Tu·GTP leads to a further change in the mode of attachment of the nucleotide to EF-Tu. As revealed from our experiments the ribose moiety of the guanosine nucleotides is relatively freely accessible on the surface of the EF-Tu·GDP or EF-Tu·GTP complexes. It follows that either a conformation of the ribose ring or the structure of the protein in the vicinity of the guanosine nucleotides are considerably different in both complexes.

The aminoacyl residue is especially important for the recognition of aminoacyl-tRNA by EF-Tu·GTP. Based on our experiments with modified tRNAs, we suggest that the elongation factor Tu requires a protonated α-amino group of the amino acid for the recognition. The aminoacyl residue is attached during this interaction probably in an orthoacid form to both 2' and 3' hydroxyl groups of the terminal adenosine. In the ternary complex the protonated α-amino group is not involved in the formation of a salt bridge with the 5' phosphate of the tRNA.

The nucleobases of the CCA-end of the aminoacyl-tRNA probably do not interact directly with EF-Tu, since base substitutions on positions 74 or 75 do not impair the ternary complex formation. Thus, the bases of the 3' terminus of aminoacyl-tRNA may be capable of an interaction with the peptidyl-transferase A-site after the ternary complex binds to ribosomes. The attachment of a fluorescence label to cytidine 75 does not impair the ternary complex formation with this modified tRNA. This allows a facile spectrophotometric determination of equilibrium parameters in ternary complexes. The affinity chromatography of aminoacyl-tRNAs on immobilized EF-Tu·GTP and the fluorescence-spectroscopic measurements are thus useful for the quantitative studies of the mechanism and fidelity of translation.

References

Adkins HJ, Miller DL, Johnson AE (1983) Biochemistry 22: 1208-1217
Arai K, Maeda T, Kawakita M, Ohnishi S, Kaziro Y (1976) J Biochem (Tokyo) 80: 1047-1055
Arai K, Arai T, Kawakita M, Kaziro Y (1975) J Biochem (Tokyo) 77: 1095-1106
Arai K, Arai T, Kawakita M, Kaziro Y (1977) J Biochem (Tokyo) 81: 1335-1346
Arai K, Arai N, Nakamura S, Oshima T, Kaziro Y (1978) Eur J Biochem 92: 521-531
Azhayev AV, Ozols AM, Bushnev AS et al. (1979) Nucl Acids Res 6: 625-643
Beres L, Lucas-Lenard J (1973) Biochemistry 12: 3998-4002
Bhuta P, Kumar G, Chladek S (1982) Biochemistry 21: 899-905
Boutorin AS, Clark BFC, Ebel JP, Kruse TB, Petersen HU, Remy P, Vassilenko S (1981) J Mol Biol 152: 593-609
Calendar R, Berg P (1966) Biochemistry 5: 1690-1695
Crane LJ, Miller DL (1974) Biochemistry 13: 933-939
Derwenskus KH, Sprinzl M (1983) FEBS Lett 151: 143-147
Douglass J, Blumenthal T (1979) J Biol Chem 254: 5383-5387
Duisterwinkel FJ, Kraal B, De Graaf JM et al. (1984) EMBO J 3: 113-120
Eccleston JF (1981) Biochemistry 20: 6265-6272
Fischer W, Derwenskus KH, Sprinzl M (1982) Eur J Biochem 125: 143-149
Hay RW, Porter LJ (1967) J Chem Soc (B) 1261-1264
Hecht SM, Tan KH, Chinault C, Arcari P (1977) Proc Natl Acad Sci USA 74: 437-441
Jonak J, Smrt J, Holy A, Rychlik L (1980) Eur J Biochem 105: 315-320
Kawakita M, Arai K, Kaziro Y (1974) J Biochem (Tokyo) 76: 801-809
Kaziro Y (1978) Biochim Biophys Acta 505: 95-127
Konigsberg W, Godson GN (1983) Proc Natl Acad Sci USA 80: 687-691
Kruse TA, Clark BFC, Sprinzl M (1978) Nucleic Acids Res 5: 879-892
Miller DL and Weissbach H (1977) In: Weissbach H, Pestka S (eds) Molecular Mechanisms of Protein Biosynthesis. Academic, New York, pp 323-373
Nakano A, Miyazawa T, Nakamura S, Kaziro Y (1980) Biochemistry 19: 2209-2215
Ohta S, Nakanishi M, Tsuboi M, Arai K, Kaziro Y (1977) Eur J Biochem 78: 599-608
Parlato G, Pizzano R, Picone D, Guesnet J, Fasano O, Parmeggiani A (1983) J Biol Chem 258: 995-1000
Picone D, Parmeggiani A (1983) Biochemistry 22: 4400-4405
Pingoud A, Kownatzki R, Maass G (1976) Nucl Acids Res 4: 327-338
Pingoud A, Urbanke C (1980) Biochemistry 19: 2108-2112
Printz MP, Miller DL (1973) Biochem Biophys Res Commun 53: 149-156
Ravel JM, Shorey RL, Shive W (1967) Biochem Biophys Res Commun 29: 68-73
Schulman LH, Pelka H, Sundari RM (1974) J Biol Chem 249: 7102-7110
Sedlacek J, Jonak J, Rychlik I (1976) FEBS Lett 68: 208-210

Sprinzl M, Kucharzewski M, Hobbs JB, Cramer F (1977) Eur J Biochem 78: 55-61
Sprinzl M, Faulhammer HG (1978) Nucleic Acids Res 5: 4837-4853
Sprinzl M, Siboska GE, Pedersen JA (1978) Nucleic Acids Res 5: 861-877
Sprinzl M, Graeser E (1980) Nucleic Acids Res 8: 4737-4744
Taiji M, Yokoyama S, Miyazawa T (1983) Biochemistry 22: 3220-3225
Tanada S, Kawakami M, Nishio K, Takemura S (1982) J Biochem (Tokyo) 91: 291-299
Wagner T, Sprinzl M (1980) Eur J Biochem 108: 213-221
Wikman FP, Siboska GE, Petersen HU, Clark BFC (1982) EMBO J 1: 1095-1100
Wilson GE, Cohn M, Miller DL (1978) J Biol Chem 253: 5764-5768
Wittinghofer A, Warren WF, Leberman R (1977) FEBS Lett 75: 241-243
Yamane T, Miller DL, Hopfield JJ (1981) Biochemistry 20: 7059-7064

Transfer RNA Binding to Ribosomes

tRNA Binding to and Topographical Arrangement on *E.coli* Ribosomes

W. WINTERMEYER, R. LILL, H. PAULSEN, and J.M. ROBERTSON[1]

1 Introduction

In order to obtain an understanding of the molecular mechanism of ribosomal protein synthesis, the large body of biochemical data on the participating molecules must be complemented by structural, thermodynamic, and kinetic information. Typically, this entails the use of physicochemical methods. The present contribution illustrates how fluorescence techniques can be used to study the mechanism of tRNA-ribosome complex formation and dissociation as well as the topographical arrangement of the tRNAs in the ribosomal A and P sites. The chapter starts with the biochemical characterization of the ribosomes with respect to tRNA binding.

2 Ribosome Activity and tRNA Binding Sites

The generally accepted model of the ribosomal elongation cycle is based upon the existence of two tRNA binding sites on the ribosome: one for peptidyl-tRNA (P site) and another for aminoacyl-tRNA (A site). From the P site-bound peptidyl-tRNA, the peptide is transferred to the aminoacyl-tRNA to yield peptidyl-tRNA in the A site and deacylated tRNA in the P site. During the subsequent translocation step, the A site-bound peptidyl-tRNA is translocated to the P site, and the deacylated tRNA is released.

Additional tRNA binding sites other than the two canonical ones have been discussed repeatedly. For instance, an entry (Hardesty et al. 1969) or recognition (Lake 1977) site has been proposed, from which the aminoacyl-tRNA reaches the A site only after GTP hydrolysis and codon recognition have taken place. Since at least the anticodon region of the tRNA in both recognition and A sites is bound to the same site on the ribosome, such a recognition site does not constitute an independent, non-overlapping site.

The existence of a third, independent site on *E. coli* ribosomes being accessible only for deacylated tRNA has been reported by Rheinberger et al. (1981). Elaborating a proposal of Noll (1966), this site has been discussed to function as an exit site (E site) which during translocation receives the deacylated tRNA from the P site (Rheinberger et al. 1981), and from which it is released only after occupation of the A site (Rheinberger and Nierhaus 1983). Except for one recent report (Grajewskaja et al. 1982), the

---

[1] Institut für Physiologische Chemie, Physikalische Biochemie und Zellbiologie der Universität München, Goethestr. 33, D-8000 Munich 2, FRG

Mechanisms of Protein Synthesis
ed. by Bermek
© Springer-Verlag Berlin Heidelberg 1984

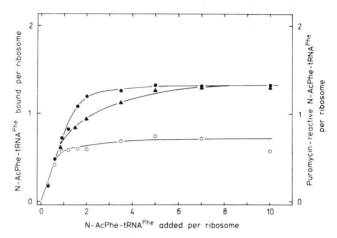

Fig. 1. Titration of ribosomes with N-AcPhe-tRNA^{Phe}. Increasing amounts of N-Ac(^{14}C)Phe-tRNA^{Phe} from yeast (1.5 nmol N-AcPhe/A_{260} unit; 504 Ci/mol) were incubated with 0.25 μM E. coli ribosomes (prepared as described by Robertson and Wintermeyer 1981) and poly(U) (1 A_{260} unit ml) for 30 min (20 mM Mg^{2+}) or 90 min (10 mM Mg^{2+}) at 25°C. Ribosome binding (●▲) and puromycin reactively (○) were measured as described in the reference. (●○) Buffer A (20 mM magnesium acetate, 50 mM Tris/HCl, pH 7.5, 90 mM NH$_4$ Cl, 40 mM KCl, 1 mM dithioerythritol); (▲) buffer B (as buffer A, but 10 mM magnesium acetate)

existence of the third site has been doubted because the additional binding could not be detected in equilibrium centrifugation experiments (Schmitt et al. 1982) and because stoichiometric release of deacylated tRNA was observed within one round of translocation in a column-bound poly(U)-ribosome system (Spirin 1983).

As an important prerequisite for the interpretation of our physicochemical data on thermodynamics and kinetics of tRNA binding to ribosomes, we have studied the properties of E. coli tight-couple ribosomes with respect to tRNA binding in some detail. Saturation titration experiments were performed with the peptidyl-tRNA analogue N-AcPhe-tRNA^{Phe} from both yeast (Fig. 1) and E. coli. It was found that at Mg^{2+} concentrations between 10 and 20 mM up to 1.3 molecules of N-AcPhe-tRNA^{Phe} were bound per ribosome (calculated on the basis of 23 pmol/A_{260} unit), which according to the puromycin reactivity were equally distributed between P and A sites. The same result was obtained with N-AcPhe-tRNA^{Phe} from E. coli. We conclude that the ribosome preparation used for the experiment was 65% active in binding N-AcPhe-tRNA^{Phe} to both P and A sites simultaneously. This result is in accordance with the results of Kirillov and Semenkov (1982) and contrasts with the exclusion of N-AcPhe-tRNA^{Phe} from A site binding by P site-bound N-AcPhe-tRNA^{Phe}, as postulated by Rheinberger et al. (1981).

The tRNA binding capacity of the ribosomes was also measured by a competition binding assay, in which the ribosomes were first incubated with increasing amounts of N-AcPhe-tRNA^{Phe} of low specific radioactivity or of deacylated tRNA^{Phe} to fill all accessible sites successively. Unoccupied P sites could then be determined from the puromycin reactivity of subsequently added N-AcPhe-tRNA^{Phe} of high specific radioactivity as an indicator, unoccupied A sites by the binding of the ternary complex of

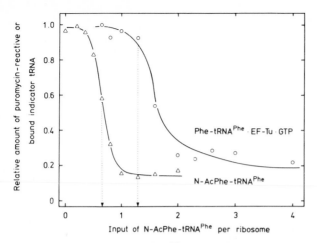

Fig. 2. Binding of N-Ac-Phe-tRNA^{Phe} into the ribosomal P and A sites as followed by subsequent indicator reactions. Ribosomes were incubated with increasing amounts of N-AcPhe-tRNA^{Phe} of low specific radioactivity (10 Ci/mol; charging level around 1.5 nmol/A$_{260}$ unit) as in Fig. 1 (buffer A). In order to determine remaining free P sites, aliquots of the reaction mixtures were incubated for 20 min at 25°C with subsequently added N-AcPhe-tRNA^{Phe} of high specific radioactivity (504 Ci/mol), and the puromycin reactivity was measured (△). For the determination of free A sites, the amount of bound ternary complex Phe-tRNA^{Phe}/EF-Tu/GTP (high specific radioactivity, 504 Ci/ mol), which had been added to another set of aliquots and incubated for 30 s at 25°C, was measured by the filtration assay (○). The continuous lines were obtained by least-squares fitting assuming a sequential binding model. Using known binding constants for N-AcPhe-tRNA^{Phe} and for the ternary complex (Lill, unpublished data), the variable parameter obtained from the fit was the concentration of P or A sites being active in binding N-AcPhe-tRNA(Phe (*dotted lines*)

Phe-tRNA^{Phe} with elongation factor Tu and GTP. Sigmoidal competition titration curves were obtained (Fig. 2), from which the concentration of active tRNA binding sites was determined to about 65% of the total ribosome concentration, the same estimate as obtained by the saturation titration. Interestingly, the same activity, 68%, was found in polyphenylalanine synthesis, as measured by the chain termination assay with radioactively labeled puromycin (Chinali and Parmeggiani 1980). This observation indicates that the active fraction of the ribosomes performs all functions in elongation with about the same efficiency.

Ribosome binding of deacylated tRNA was measured with tRNA^{Phe} from both yeast and *E. coli*, the 5' end of which had been labeled with ^{32}P-phosphate (Fig. 3). In both cases, saturation was reached when three tRNA^{Phe} molecules were bound per active ribosome. The activity, 65%, again was determined by the indicator titrations; in addition, by the indicator titrations the site location of the deacylated tRNA^{Phe} could be determined unambiguously. Interestingly, binding of the third tRNA^{Phe} from yeast could not be detected when the reaction mixture was diluted with buffer prior to the filtration (Fig. 3), indicating a lower stability of the heterologous E site complex as compared to the homologous one. The different affinity of the E site for yeast and *E. coli* tRNA^{Phe} is also reflected in the stoichiometry of A site binding. With yeast tRNA^{Phe} the A site is filled with a stoichiometry of two, i.e., the A site is the second site to be filled after the P site. In contrast, with *E. coli* tRNA^{Phe} an A site binding

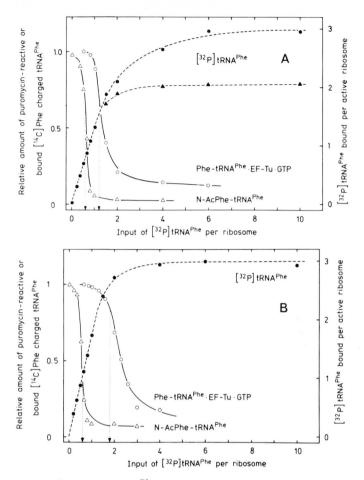

Fig. 3 A-B. Ribosome binding of ^{32}P-labeled tRNA^Phe from yeast (**A**) and *E. coli* (**B**). Increasing amounts of tRNA^Phe labeled with ^{32}P-phosphate at the 5'end, were added to ribosomes as in Fig.1 (buffer A). The indicator reactions were performed as in Fig. 2. For the determination of the binding level, the reaction mixtures were applied to the nitrocellulose filters either directly (•) or after dilution with cold buffer A (▲)

stoichiometry of three was observed, indicating that in this case the E site is filled prior to or concomitantly with the A site.

The stability of the E site complexes exhibited a rather strong Mg^{2+} dependence. Thus, the binding of yeast tRNA^Phe to the E site is hardly detectable anymore at 10 mM Mg^{2+} (not shown). Similarly, *E. coli* tRNA^Phe at 10 mM Mg^{2+} is chaseable from the E site by the addition of excess tRNA^Phe (Fig. 4). Consequently, at Mg^{2+} concentrations of 10 mM and below, at which our system works optimally in polyphenylalanine synthesis (8 mM), only a fraction of the tRNA from the P site after translocation remains bound to the E site (Robertson et al., unpublished results). On the basis of these results, it appears unlikely that the E site-bound tRNA contributes appreciably to the fixation of the mRNA, as has been postulated by Rheinberger et al.(1981). Furthermore, the re-

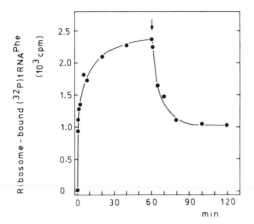

Fig. 4. Dissociation of ^{32}P-labeled *E. coli* tRNA^{Phe} from the E site. To P site-blocked, poly(U)-programed ribosomes ^{32}P-labeled tRNA^{Phe} from *E. coli* was added and the successive binding of the tRNA to E and A sites was followed by the filtration assay. At the time marked by the *arrow*, an eightfold excess of unlabeled tRNA^{Phe} was added. The decrease corresponds to the amount of E site-bound labeled tRNA, the final plateau represents the A site-bound tRNA

sults show that the release of the tRNA from the E site does not depend upon A site occupancy, as has been postulated recently (Rheinberger and Nierhaus 1983).

3 Kinetics of tRNA-ribosome Complex Formation

Protein synthesis in *E. coli* proceeds at a rate of approximately 15 amino acids incorporated into protein per ribosome per second, implying elementary steps of the elongation cycle in the range of milliseconds (Gouy and Grantham 1980). Thus, in order to measure the rates of individual steps and to isolate short-lived intermediates of ribosomal protein synthesis, fast kinetic techniques have to be applied. We have been using the fluorescence stopped-flow technique for studying the binding of fluorescent derivatives of yeast tRNA^{Phe} to *E. coli* ribosomes. As a fluorescence marker, proflavine was covalently incorporated into either the anticodon loop (tRNA$^{Phe}_{Prf37}$) or into the D loop (tRNA$^{Phe}_{Prf16/17}$) of tRNA^{Phe} by replacing the odd bases at positions 37 and 16/17, respectively (Wintermeyer and Zachau 1979).

Previously, we have utilized the label in the anticodon loop to study the binding of tRNA^{Phe} to poly(U)-programed, vacant ribosomes (Wintermeyer and Robertson 1982). In this system, two kinetic steps accompanying P site binding are observed which can be characterized as a recombination step followed by a slow rearrangement. Upon blocking the P site, the slow step is repressed, whereas the recombination step remains essentially unchanged; this indicates that the latter reflects the binding of the tRNA to some intermediate site.

The steady-state binding experiments have revealed three ribosomal binding sites for deacylated tRNA^{Phe} which, in the case of yeast tRNA^{Phe}, are filled in the order P, A, E, as judged from the equilibrium data. However, as mentioned above, it is possible that the tRNA, before reaching the P or A sites, is transiently bound to another site, the E site, for instance. In order to study this problem, we have utilized the D loop-labeled tRNA^{Phe} derivative for stopped-flow experiments with poly(U)-programed ribosomes (Fig. 5). The same two steps leading to the P site complex, as described above,

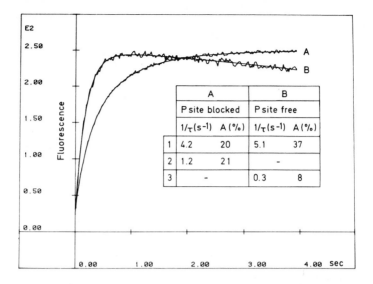

Fig. 5. Kinetics of tRNA$^{Phe}_{Prf16/17}$ binding to vacant (**B**) and P site-blocked (**A**) poly(U)-programed ribosomes. The stopped-flow experiments were performed by rapidly mixing equal volumes of the tRNA solution (final concentration after mixing 0.01 μM) and of the ribosome solution (final concentration 0.1 μM), which in **B** had been preincubated with a stoichiometric amount of unmodified tRNA^{Phe} in order to block the P site. The experiments were performed at 20°C in buffer C (50 mM Tris/HCl, pH 7.6, 25 mM KCl, 70 mM NH$_4$Cl, 18 mM magnesium acetate, 3 mM 2-mercapto ethanol). The data were evaluated by two-exponential least-squares parameter fitting, and the functions obtained are given by the noise-less lines

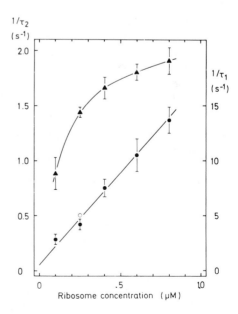

Fig. 6. Dependence of the relaxations of Fig. 5 upon the concentration of ribosomes. Stopped-flow experiments were performed with increasing amounts of P site-blocked ribosomes as in Fig. 5

were observed with vacant ribosomes. Interestingly, with P site-blocked ribosomes an additional fluorescence change was observed, which exhibited an intermediate rate, around 1 s^{-1}, whereas the slow rearrangement step was suppressed.

The two steps of Fig. 5 were characterized further by their dependence upon the concentration of ribosomes (Fig. 6). For the fast step, the same linear dependence as with vacant ribosomes was observed, indicating it represents the recombination step described before. The saturating behavior of the intermediate step suggests a rearrangement which is coupled to the binding step, as illustrated in scheme 1 (R denotes P site-blocked, poly(U)-programed ribosomes, T the fluorescent tRNA derivative):

$$R+T \underset{k_{21}}{\overset{k_{12}}{\rightleftharpoons}} RT' \underset{k_{32}}{\overset{k_{23}}{\rightleftharpoons}} RT'' \tag{1}$$

It should be mentioned that from the kinetic data the alternative or parallel pathway leading directly to RT" cannot be excluded. Therefore, the sequential pathway, the rate constants of which are evaluated in the following, should be considered a minimal model. The four rate constants of the two coupled steps of scheme 1 were evaluated from linear plots of the sums and the products of the reciprocal relaxation times, given in Fig. 6, versus the concentration of ribosomes (Bernasconi 1976): $k_{12} = (17\pm3)$ $(\mu M*s)^{-1}$, $k_{21} = 0.3\pm0.1$ s^{-1}, $k_{23} = 1.8\pm0.2$ s^{-1}, $k_{32} = 0.1\pm0.1$ s^{-1}. Generally, because of the relatively high stability of the complexes, the off constants intrinsically contain large uncertainties.

In order to determine the binding site(s) of the fluorescent tRNA in the intermediate complexes RT' and RT" (scheme 1), the stopped-flow experiments were repeated with ribosomes which had been saturated with a tenfold excess of yeast tRNA^{Phe}. In these ribosomes, all three sites are occupied (cf. Fig. 2), and the subsequent binding of fluorescent tRNA can occur only via exchange with E site-bound tRNA, because both P and A site-bound tRNAs exchange extremely slowly (half times of hours). In such an experiment, the rate of E site binding is expected to be limited by the rate of dissociation of the prebound tRNA from the E site. In fact, the two steps of Fig. 5 are still observed under these conditions, although at slower rates: 1.5 and 0.3 s^{-1}. This may be interpreted to indicate that in both intermediate complexes, RT' and RT", the tRNA is bound to the E site.

The above interpretation implies that the transition from RT' to RT" involves a conformational change of the tRNA or the ribosome or both. Alternatively, one may assume that the tRNA occupies different binding sites in RT' and RT". The latter alternative appears somewhat less likely since it requires a fourth independent binding site, which has not been observed in the titration experiments. It cannot be excluded, however, because the rapidly dissociating complex RT' ($k_{21} = 0.3$ s^{-1}), may not be detectable by filtration.

In the physiological situation, the dissociation of deacylated tRNA from the E site during translocation is the important step, rather than binding to the E site. In principle, the rate constant of dissociation can be determined from the concentration dependence of the reciprocal relaxation times (Fig. 6). As mentioned above, the values obtained are rather inaccurate because the ordinate intercepts are small and uncertain. We, therefore, tried to determine the dissociation rate constants more directly. Chase experiments

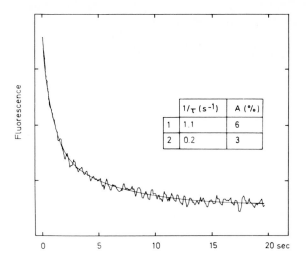

Fig. 7. Dissociation of tRNA$^{Phe}_{Prf16/17}$ from the ribosomal E site. See text. Experimental conditions as in Fig. 5

have been performed, in which fluorescent deacylated tRNA^{Phe} was bound to the E site by exchange for prebound non-labeled tRNA^{Phe}. The dissociation of the fluorescent tRNA from the E site was initiated by rapidly mixing the complex with a tenfold excess of nonlabeled tRNA^{Phe} (Fig. 7). The fluorescence signal decreased in a biphasic manner, the apparent rate constants being about 1 and 0.2 s^{-1}. These figures are similar to those found for the rearrangement of RT' to RT" (1.8 s^{-1}, Fig. 6) and the dissociation of RT' (0.3 s^{-1}). This similarity is expected from the sequential model. The remaining differences most likely are due to the influence of ribosome binding on non-labeled tRNA^{Phe}, which is present in large excess in the chase experiment.

An interesting implication of the rate constants obtained for scheme 1 should be discussed. RT" is formed much faster (k_{23} = 1.8 s^{-1}) than the P site complex (k = 0.3 s^{-1}; Wintermeyer and Robertson 1982). Inspite of this difference, the formation of RT" was not observed with vacant ribosomes. This finding indicates that tRNA binding to the E site depends upon occupation of the P site. The same conclusion can be drawn from the titration experiments of Fig. 2, which showed that the P site is the first site to be filled. This raises the possibility that the P site-bound tRNA^{Phe} exerts an allosteric effect on the conformation of the ribosome, which influences the accessibility or stability of the E site.

4 Topographical Arrangement of A and P Site-bound tRNAs

A prerequisite for understanding the mechanism of protein synthesis on a molecular level is detailed knowledge of the topography of the ribosomal complex containing the tRNAs and the mRNA. Of central interest is the arrangement of the two tRNA molecules that are present simultaneously on the ribosome during important steps of the elongation cycle. Several models have been proposed for the functioning of tRNA within the ribosomal complex (Woese 1970; Rich 1974; Lake 1977), which imply certain orientations of the two tRNA molecules relative to each other.

In order to define the relative orientation of the two ribosome-bound tRNAs, measurements of singlet-singlet energy transfer have proven useful. The basis of the method is the dependence of the transfer efficiency upon the distance between the energy donor and acceptor. Theoretically, the method is suitable to measure intra- or intermolecular distances up to say 80 Å precisely. In practice it is limited, however, because the relative orientation of the transition dipoles of the interacting chromophores, which also influences the transfer efficiency, in most cases is not known exactly. The range of uncertainty can be limited on the basis of fluorescence anisotropy data, which reflect the mobility of the chromophores (Dale and Eisinger 1974), thus leading to more or less narrow ranges of possible distances for a given donor-acceptor separation.

Craig et al. (1978) and Fairclough and Cantor (1979) determined the distance between the adjacent anticodon loops by measuring the energy transfer between wybutine, a natural fluorophor present 3' to the anticodon in yeast tRNA^{Phe}, and proflavine in place of the wybutine (Wintermeyer and Zachau 1979) in the second tRNA. The anticodon loop-anticodon loop separation found was rather short (less than either 30 or 20 Å, respectively). A large separation of the midsections of the two tRNA molecules was excluded by Johnson et al. (1982) after measuring the energy transfer between ribosome-bound *E. coli* tRNAs labeled at 4-thiouridine. The 3' termini of the two tRNAs most likely are rather close to each other since the aminoacyl ends must come into contact for peptidyl transfer.

The results summarized so far help to exclude some possible arrangements of the two tRNAs, but are insufficient to support the choice of one particular orientation. Therefore, we determined four tRNA-tRNA distances in the ribosomal complex, including the separations between the anticodon loops and between the D loops, each by several independent energy transfer measurements. The latter distances, particularly that between the two D loops, are critical for the evaluation of possible arrangements of the two tRNAs in the ribosomal A and P sites.

Three fluorophores were used. The energy-donating groups were the natural fluorophor wybutine in the anticodon loop and proflavine replacing it as well as proflavine in the D loop. Proflavine (wybutine donor) or ethidium (proflavine donor) in both anticodon and D loops were used as the energy accepting groups. tRNA^{Phe} carrying ethidium in the anticodon loop can be isolated in two isomeric species with different spectroscopic properties: tRNA$^{Phe}_{Etd37B}$ and tRNA$^{Phe}_{Etd37C}$ (Wintermeyer and Zachau 1979); both forms were used here. Thus, we were able to measure each distance (except for that between the D loops) with at least two different donor-acceptor pairs. Furthermore, all measurements were performed both ways, by exchanging the respective location of donor and acceptor-labeled tRNA in the two ribosomal sites. The ribosomes carrying the N-AcPhe-tRNA^{Phe} or tRNA^{Phe} derivatives in both A and P sites were isolated by centrifugation in order to eliminate unbound fluorescent tRNA. This procedure also removed tRNA from the E site. The site location of N-AcPhe-tRNA^{Phe} in all complexes was determined by the puromycin assay.

Two spectroscopic effects can be seen upon energy transfer either from wybutine to proflavine (Fig. 8) or from proflavine to ethidium (Fig. 9), which are shown as examples. The donor fluorescence is quenched in the presence of the acceptor, while the acceptor emission is stimulated by exciting the donor, the behavior expected for singlet-singlet energy transfer. The transfer efficiency can be determined by measuring either

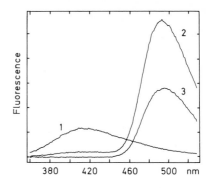

Fig. 8. Fluorescence energy transfer from A site-bound N-AcPhe-tRNA^{Phe} to P site-bound tRNA$^{Phe}_{Prf37}$. The ribosome complexes were prepared in buffer A and isolated by ultracentrifugation. The samples contained N-AcPhe-tRNA^{Phe} plus *E. coli* tRNA^{Phe} (spectrum 1, donor alone), N-AcPhe-tRNA^{Phe} plus tRNA$^{Phe}_{Prf37}$ (spectrum 2, donor plus acceptor), and *E. coli* tRNA^{Phe} plus tRNA$^{Phe}_{Prf37}$ (spectrum 3, acceptor alone) in the A and P sites, respectively. According to the nitrocellulose filtration and the puromycin assays, 96% of the N-AcPhe-tRNA^{Phe} were bound and 3% of the bound material puromycin reactive. Corrected transfer efficiency 0.88. Figure reproduced from Paulsen et al. (1983)

Fig. 9. Fluorescence energy transfer from A site-bound N-Ac-Phe-tRNA$^{Phe}_{Prf16/17}$ to P site-bound tRNA$^{Phe}_{Etd16/17}$. The three samples were prepared as in Fig. 8. Corrected transfer efficiency 0.25. Figure reproduced from Paulsen et al. (1983)

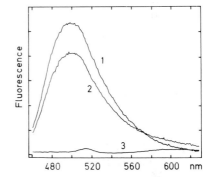

the quenching of the donor fluorescence or the stimulation of the acceptor fluorescence. The data given below have been obtained from the donor spectra.

The efficiencies of energy transfer between the various ribosome-bound donor-acceptor pairs are given in Table 1, as are the ciritcal distances, R_0, used to calculate the distance ranges given in the table. At first sight, the ranges of possible distances given in Table 1 appear rather broad. However, because there are at least two sets of data for each distance, these ranges can be narrowed significantly. Since the true distance lies within each one of the distance ranges obtained from the various donor-acceptor pairs, it has to lie within the range in which all the measurements overlap. For instance, the six distance ranges resulting from measurements of the anticodon loop-anticodon loop separation are shifted against each other, but overlap in the region between 20 and 27 Å. Therefore, a distance range of 24±4 Å is given as the overall result of these measurements in the last column of Table 1. For the remaining three distances, the ranges obtained from various donor-acceptor pairs are largely congruent. Therefore,

Table 1. Transfer efficiencies and donor-acceptor separations[a]

Distance	Donor (site)[b]	Acceptor[c]	Transfer efficiency	R_o ranges (Å)	Distance ranges (Å)	Donor-Acceptor separation (Å)
Anticodon loop-anticodon loop	Wye37[d](A)	Prf37	0.88	24-39	22±5	
	Prf37[d](A)	Etd37C	0.75	22-46	28±10	
	Prf37[d](A)	Etd37B	0.91	22-47	23±9	24±4
	Wye37[d](P)	Prf37	0.62	24-38	27±7	
	Prf37[c](P)	Etd37C	0.68	21-43	28±10	
	Prf37[c](P)	Etd37B	0.70	21-43	28±10	
D loop-D loop	Prf16/17[d](A)	Etd16/17	0.30	21-38	34±11	
	Prf16/17[d](P)	Etd16/17	0.31	22-38	34±10	35±9
	Prf16/17[c](P)	Etd16/17	0.24	22-39	37±11	
Anticodon loop (Asite)-D loop	Wye37[d](A)	Prf16/17	0.15	25-40	46±12	
	Prf37[d](A)	Etd16/17	0.12	24-46	50±16	46±12
(P site)	Prf16/17[c](P)	Etd37C	0.11	21-40	44±14	
	Prf16/17[c](P)	Etd37B	0.11	21-40	44±15	
Anticodon loop (P site)-D loop (A site)	Prf16/17[d](A)	Etd37C	0.19	19-38	36±13	
	Prf16/17[d](A)	Etd37B	0.18	19-39	37±13	38±10
	Wye37[d](P)	Prf16/17	0.32	25-39	38±10	
	Prf37[c](P)	Etd16/17	0.27	22-43	38±12	

[a]The energy transfer efficiencies were determined from the donor spectra; values are averaged from at least two experiments, the maximum variation being 0.02. The ranges for the critical distances, R_o, were determined from anisotropy data according to Dale and Eisinger (1974). The distance ranges given take into account the ranges of R_o and the experimental error

[b]Wye, wybutine; Prf, proflavine; Etd, ethidium

[c]Deacylated tRNA[Phe]

[d]N-AcPhe-tRNA[Phe]

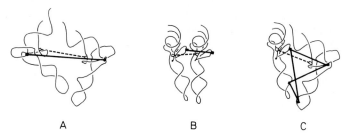

Fig. 10 A-C. Possible orientations of the two tRNAs bound to the ribosomal A and P sites. In all three depictions, the P site-bound tRNA is drawn to the *left*. The D loop-D loop distances are indicated by *continuous lines* and the separation between fluorophores attached to 4-thiouridine 8, measured by Johnson et al. (1982), by *broken lines*. In C the other distances measured here are also indicated. Figure reproduced from Paulsen et al. (1983)

the overall results, again represented by the overlap interval of the homologous distance ranges, retain a rather wide range of uncertainty.

The value for the anticodon loop-anticodon loop separation obtained from our measurements, 24±4 Å, is comparable with published values determined by the energy transfer measurements discussed above. Thus, all the data indicate that the anticodon loops are rather close to each other, allowing for simultaneous binding of the anticodons to contiguous codons.

The other distances given in Table 1 place some restrictions on possible models of the tRNA-ribosome complex. With the assumption that not only the anticodon loops, but also the acceptor ends, are close to each other, three basic arrangements have to be considered (Fig. 10). It can be seen that the possible orientations are largely determined by the separation of the tRNA midsections. The most extended arrangement of the two tRNAs separates the D loops by more than 60 Å (Fig. 10A), while a distance of less than 25 Å would be estimated for the side-by-side arrangement (Fig. 10B). Thus, the distance between the two D loops that we have measured, 35±9 Å, suggests an orientation of the two tRNAs between these two extreme possibilities, such as that depicted in Fig. 10 C. Such an arrangement, with the planes of the two tRNAs perpendicular to each other was already considered by Rich (1974). On the basis of tRNA-ribosome cross-linking data, Ofengand (1980) favored a similar model with an angle of about 45° between the planes of the two tRNA molecules. Our data seem to be compatible with angles in that range, 60±30°.

Although Fig. 10C or related arrangements accomodate our data rather well, we should like to emphasize that this model is meant to illustrate general features rather than give a detailed structural picture of tRNA-ribosome complexes. Probably the most severe limitation of all such models is the assumption that the conformation of the ribosome-bound tRNA is represented by the structure prevailing in the crystal. There are data obtained by several techniques indicating that the tRNA conformation is changed by ribosome binding, the extent of the change being undefined (Farber and Cantor 1980; Robertson and Wintermeyer 1981; Peattie and Herr 1981). One such possible conformational transitions are better defined, the knowledge of the distances between various parts of the two ribosome-bound tRNA molecules may aid the construction of more detailed models of the arrangement of the tRNAs in the ribosomal binding sites.

Acknowledgement. We thank H.G. Zachau for his interest and encouragement. One of us (H.P.) received a grant from the Studienstiftung des Deutschen Volkes, another (W.W.) a Heisenberg Fellowship from the Deutsche Forschungsgemeinschaft. The work was supported by the Deutsche Forschungsgemeinschaft and the Fonds der Chemischen Industrie.

References

Bernasconi CF (1976) Relaxation Kinetics. Academic, New York

Chinali G, Parmeggiani A (1980) The coupling with polypeptide synthesis of the GTPase activity dependent on elongation factor G. J Biol Chem 255: 7455-7459

Craig BB, Odom OW, Foyt D, White JM, Hardesty B (1978) Fluorescence polarization and energy transfer studies of tRNAs bound into the ribosomal donor and acceptor sites.

In: Agris PF (ed) Biomolecular Structure and Function, Academic, New York, pp 545-553

Dale RE, Eisinger J (1974) Intramolecular distances determined by energy transfer. Dependence on orientational freedom of donor and acceptor. Biopolymers 13: 1573-1605

Fairclough RH, Cantor CR (1979) The distance between the anticodon loops of two tRNAs bound to the 70S *Escherichia coli* ribosome. J Mol Biol 132: 575-586

Farber N, Cantor CR (1980) Comparison of the structures of free and ribosome-bound transfer RNA^{Phe} by using slow tritium exchange. Proc Natl Acad Sci USA 77: 5135-5139

Gouy M, Grantham R (1980) Polypeptide elongation and tRNA cycling in *Escherichia coli*: a dynamic approach. FEBS Lett 115: 151-155

Grajevskaja RA, Ivanov YV, Saminsky EM (1982) 70S ribosomes of *Escherichia coli* have an additional site for deacylated tRNA binding. Eur J Biochem 128: 47-52

Hardesty B, Culp W, McKeehan W (1969) The sequence of reactions leading to the synthesis of a peptide bound on reticulocyte ribosomes. Cold Spring Harbor Symp Quant Biol 34: 331-345

Johnson AE, Adkins HJ, Matthews AE, Cantor CR (1982) Distance moved by transfer RNA during translocation from the A site to the P site on the ribosome. J Mol Biol 156: 113-140

Kirillov SV, Semenkov YP (1983) Non-exclusion principle of AcPhe-tRNA^{Phe} interaction with the donor and acceptor sites of *Escherichia coli* ribosomes. FEBS Lett 148: 235-238

Lake JA (1977) Aminocyl-transfer RNA binding at recognition site is the first step of the elongation cycle of protein synthesis. Proc Natl Acad Sci USA 74: 1903-1907

Noll H (1966) Chain initiation and control of protein synthesis. Science (Wash DC) 151: 1241-1245

Ofengand J (1980) The topography of tRNA binding sites on the ribosome. In: Chambliss G, Craven GR, Davies J, Davis K, Kahan L, Nomura M (eds) Ribosomes: structure, function and genetics. University Park Press, Baltimore, pp 497-529

Paulsen H, Robertson JM, Wintermeyer W (1983) Topological arrangement of two transfer RNAs on the ribosome. Fluorescence energy transfer measurements between A and P site-bound tRNA^{Phe}. J Mol Biol 167: 411-426

Peattie DA, Herr W (1981) Chemical probing of the transfer RNA – ribosome complex. Proc Natl Acad Sci USA 78: 2273-2277

Rheinberger HJ, Nierhaus K (1983) Testing an alternative model for the ribosomal peptide elongation cycle. Proc Natl Acad Sci USA 80: 4213-4217

Rheinberger HJ, Sternbach H, Nierhaus K (1981) Three tRNA binding sites on *Escherichia coli* ribosomes. Proc Natl Acad Sci USA 78: 5310-5314

Rheinberg HJ, Schilling S, Nierhaus K (1983) The ribosomal elongation cycle: tRNA binding, translocation and tRNA release. Eur J Biochem 134: 421-428

Rich A (1974) How tRNA moves through the ribosome. In: Nomura M, Tissieres A, Lengyel P (eds) Ribosomes. Cold Spring Harbor Laboratory, pp 871-884

Robertson JM, Wintermeyer W (1981) Effect of translocation ow topology and conformation of anticodon and D loops of tRNA^{Phe}. J Mol Biol 151: 57-79

Schmitt M, Neugebauer U, Bergmann C, Gassen HG, Riesner D (1982) Binding of tRNA in different functional states to *Escherichia coli* ribosomes as measured by velocity sedimentation. Eur J Biochem 127: 525-529

Spirin A (1984) this volume

Wintermeyer W, Robertson JM (1982) Transient kinetics of transfer ribonucleic acid binding to the ribosomal A and P sites: observation of a common intermediate complex. Biochemistry 21: 2246-2252

Wintermeyer W, Zachau HG (1979) Fluorescent derivatives of yeast tRNA^{Phe}. Eur J Biochem 98: 465-475

Woese C (1970) Molecular mechanics of translation: a reciprocating ratchet mechanism. Nature (Lond) 226: 817-820

Ribosomal Translocation

Studies on Structural Dynamics of the Translating Ribosome

A.S. SPIRIN and I.N. SERDYUK[1]

1 Introduction

The process of translation (elongation) on the ribosome is composed of the repeating cycles, each consisting of three successive steps: aminoacyl-tRNA binding, transpeptidation, and translocation (Watson 1964; Lipmann 1969). The translocation step includes significant intraribosomal displacements of a template and the products of the transpeptidation reaction: the release of deacylated tRNA, the transport of peptidyl-tRNA from one site to the other, and the shift of the template polynucelotide by one codon.

The question arises: Is any step of the elongation cycle, and particularly the translocation step, accompanied by mechanical alterations of the ribosomal particle? The possibility of "an alternate contraction and expansion of the ribosome" ("a pulsating ribosome contraction") in the elongation cycle was first mentioned by Lipmann *et al.* (Conway and Lipmann 1964; Nishizuka and Lipmann 1966). In the most clear form this question was posed in 1968 when the idea of a somewhat moving apart (unlocking) of the two coupled ribosomal subunits as a plausible driving mechanism for translocation was proposed (Spirin 1968).

However, the experimental answer to this question was found to be not simple. First of all, physical measurements of functioning ribosomes require *all* the particles of the sample under study to be active in elongation and present in the same functional state. This is not the case for routine ribosomal preparations where the particle population is heterogeneous and only a fraction of it manifests full activity.

2 Preparation of Active One-functional-state Ribosomes

To solve this problem, a special technique was devised for the isolation of translationally active ribosomes by using columns with poly(U) coupled to Sepharose through splittable disulfide bridges (Belitsina et al. 1975; Baranov et al. 1979). This technique allows to obtain translating ribosomes stopped either in the pretranslocation or in the posttranslocation stage of the elongation cycle and capable of continuing the cycle and elongation at any moment when substrates and proper temperature are provided.

Escherichia coli MRE-600 ribosomes washed four times with 1 M NH$_4$Cl - 0.01 M MgCl$_2$ was taken as a starting material. Poly(U) of 100 to 140 nucleotide residues long

---

[1]   Institute of Protein Research, Academy of Sciences of the USSR, Pushchino, Moscow Region, USSR

Mechanisms of Protein Synthesis
ed. by Bermek
© Springer-Verlag Berlin Heidelberg 1984

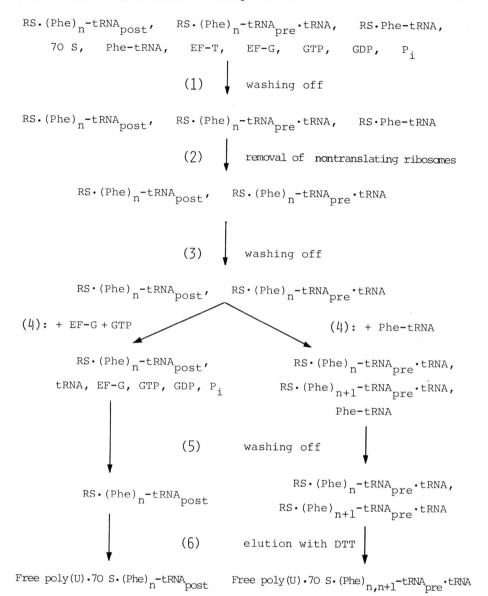

Fig. 1. Isolation of the translating ribosomes in the pretranslocation and posttranslocation states (Baranov et al. 1979; Baranov 1983. *Symbols:* RS, 70S ribosomes attached to cellulose-S-S-bound poly(U); $(Phe)_n$, polyphenylalanine with a degree of polymerization n; $(Phe)_n$-$tRNA_{post}$, polyphenylalanyl-tRNA in the posttranslocation state; $(Phe)_n$-$tRNA_{pre}$, polyphenylalanyl-tRNA in the pretranslocation state

was coupled to Sepharose through dithiodiglycolic acid residues, according to the procedure described earlier (Baranov et al. 1979). Mixture of the ribosomes and the poly(U)-Sepharose was supplemented with Phe-tRNA, elongation factors and GTP and incubated at 25°C for 10 min (in 10 mM Tris-HCl, pH 7.2., 100 mM KCl and 10 mM MgCl$_2$, without SH-compounds). The polyphenylalanine synthesis was stopped by cooling and the mixture was placed into columns. The scheme of further procedures is given in Fig. 1. (1) Columns were washed off with a high Mg^{2+} (20 mM) buffer in the cold to remove free ribosomes, tRNA, and elongation factors. (2) Then columns were washed with a buffer containing 250 mM NH$_4$Cl and 5 mM MgCl$_2$ at 37°C to remove the ribosomes bound to poly(U)-Sepharose without translation (Baranov, 1983). (3) Again the columns were washed with the cold high Mg^{2+} buffer. (4), left: To bring all the poly (U)-translating ribosomes to the posttranslocation state, the mixture of EF-G with GTP was passed through the column at 25°C. (4), right: To have all the ribosomes in the pretranslocation state, Phe-tRNA was passed through the column at high Mg^{2+} (20 mM) in the cold. (5) The columns were washed carefully from soluble and loosely bound components. (6) Posttranslocation state (left) or pretranslocation state (right) ribosomes with bound poly(U) chains were eluted from the columns with dithiothreitol.

The content of translationally active particles in the ribosome samples prepared in such a way was not less than 95% (Baranov, 1983). The fraction of the particles in a given functional state (either posttranslocation, i.e., puromycin-competent, or pretranslocation, i. e. puromycin-incompetent ones) was not less than 75%-80%.

3 Electron Microscopy

Electron microscopy images of usual nontranslating 70S ribosomes (30S·50S couples) in two main projections (Lake 1976; Vasiliev et al. 1983) are given in Fig. 2 as a standard. It is seen that the ribosomal subunits are coupled in such a manner that the head of the 30S subunit joins the "head" (central protuberance) of the 50S subunit, and the side bulge (platform) of the 30S subunit is in contact with the side protuberance (L1 ridge) of the 50S subunit. The rod-like appendage (L7/L12 stalk) can serve as a good marker for identifying the mutual orientation of the subunits. The most informative is the overlap projection where the main structural features of the two subunits and their mutual orientation can be recognized easily.

A gallery of electron micrographs of the translating 70S ribosomes in the pretranslocation (a) and posttranslocation (b) states, as well as of the nontranslating 30S·50S couples for comparison (c), is presented in Fig. 3. Analysis of the images of about 1000 particles in each case has shown that the translating 70S ribosomes in the two functional states do not differ from the nontranslating particles and from each other in respect to the mutual subunit orientation and the L7/L12 stalk position, within the resolution limits of about 20 Å (Vasiliev et al. 1983).

The conclusion can be made that transition from the pretranslocation state to the posttranslocation state results in neither a visible reorientation of the L7/L12 stalk nor a significant displacement of the subunits relative to each other, at least in the plane of the subunit interface.

Fig. 2. A gallery of electron microscopic images of standard nontranslating 70S ribosomes in two main projections. *Bottom:* photos of the 70S ribosome model in the same projections

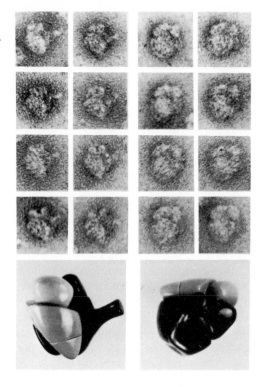

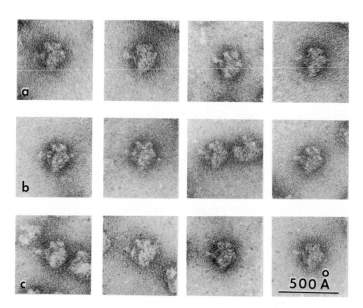

Fig. 3 a-c. A gallery of electron microscopic images of the pretranslocation state ribosomes **a**, post-translocation state ribosomes **b**, and nontranslating 30S·50S couples **c** in the overlap projection

4 Sedimentation

Earlier several groups independently reported that pretranslocation and posttranslocation state ribosomes can somewhat differ in their sedimentation coefficients (Schreier
and Noll 1971; Chuang and Simpson 1971; Gupta et al. 1971). In all the cases posttranslocation state ribosomes were slower. However, later experiments done at different
sedimentation speeds demonstrated that the difference was observed only at a high
speed (high hydrostatic pressure) and could not be detected at a lower speed (Waterson
et al. 1972). The conclusion was made that the difference in sedimentation coefficients
resulted from different stabilities of conformations of pretranslocation and posttranslocation state ribosomes to hydrostatic pressure rather than from a preexisting difference in their compactness.

It should be mentioned, however, that in all the previous experiments the so-called
pretranslocation and posttranslocation state ribosomes were not really translating (elon-
gating) particles, but complexes of 70S ribosomes with one or two aminoacyl-tRNA
derivatives in A and/or P sites.

In our experiments the preparations of pretranslocation and posttranslocation state
ribosomes at the elongation stage were obtained, as described above (Sect. 2), and used
in cosedimentation runs at different rates. Figure 4 shows that the pretranslocation
state ribosomes sedimented somewhat faster than the posttranslocation state particles.
The difference of sedimentation coefficients of about 1 S was observed at all three
speeds used (the centrifuge speeds were the same as in the early report of Lengyel's
group (Waterson et al. 1972), i.e., it is not dependent on hydrostatic pressure. This result could suggest that a difference in compactness between pretranslocation and posttranslocation state ribosomes preexists. Posttranslocation state ribosomes look slightly
less compact than ribosomes of the preceding (pretranslocation) stage. The difference
seems to be small, however; it is not surprising that it could not be revealed by electron
microscopy.

At the same time, another interpretation of the sedimentation experiments cannot
be ruled out: pretranslocation state ribosomes sediment somewhat faster because of
the presence of two tRNA residues per particle, instead of one tRNA residue in the
posttranslocation ribosome. Indeed, a tRNA residue could result in 1% increase of the
sedimentation coefficient value of the 70S particle. Additional experiments demonstrated, however, that this difference in composition did not result in different sedimentation rates under conditions of higher ionic strength (500 mM NH$_4$Cl) (Fig. 5). (The
preservation of the pretranslocation state, including retention of two tRNA residues,
under these conditions was specially checked; e.g., see Fig. 6). It seems to be not very
likely that the tRNA mass contribution was responsible for the increase of the sedimentation coefficient of the pretranslocation state ribosomes under moderate ionic strength
conditions and was not under higher ionic strength. Nevertheless, the uncertainty remains.

5 Neutron Scattering

Thus, the next problem in physical studies of functioning ribosomes arises. Indeed,
since the significant intraribosomal displacements of tRNA and mRNA molecules take

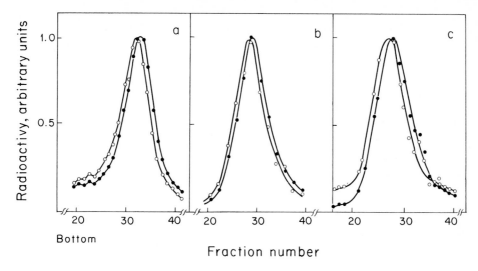

Fig. 4 a-c. Co-sedimentation of [^{14}C]pretranslocation (-○-) and [^{3}H]posttranslocation (-●-) state ribosomes in sucrose gradient containing 20 mM Tris-HCl, pH 7.5, 20 mM MgCl$_2$, 100 mM NH$_4$, Cl, 4°C. **a:** 20 000 rpm; **b:** 30 000 rpm; **c:** 40 000 rpm

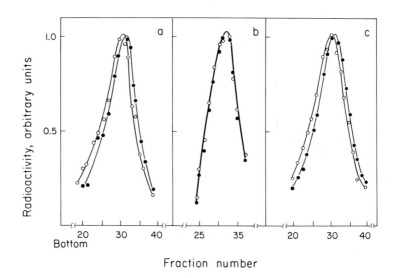

Fig. 5 a-c. Co-sedimentation of [^{14}C]pretranslocation (-○-) and [^{3}H]posttranslocation (-●-) state ribosomes in sucrose gradient at 20 000 rpm, 20 mM MgCl$_2$, 4°C. **a:** 100 mM NH$_4$Cl; **b:** 500 mM NH$_4$Cl; **c:** transferred from 500 mM NH$_4$Cl into 100 mM NH$_4$Cl

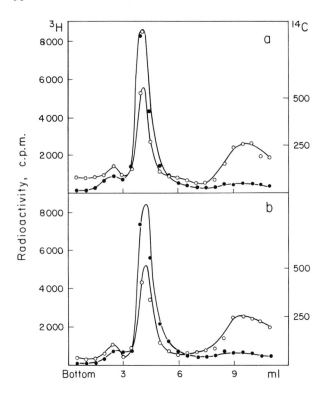

Fig. 6 a-b. Sucrose gradient sedimentation of pretranslocation state ribosomes carrying [³H]polyPhe-tRNA (-•-) in the A site and [¹⁴C]tRNA (-○-) in the P site, at 20 000 rpm, 20 mM MgCl$_2$, 4°C. **a:** 100 mM NH$_4$Cl; **b:** 500 mM NH$_4$Cl

place during the ribosomal function, the problem is to discriminate the contribution of the changes in the ligand positions from contributions of ribosomal particle alterations. This problem is especially serious: *any* total physical differences between two functional states of the ribosome, if they are detected, cannot be unambiguously interpreted since the functional states (e.g., pretranslocation and posttranslocation ones) are characterized by *different numbers* (e.g., two or one tRNA, respectively) and *different positions* of the RNA ligands on the ribosome.

For solution of this problem, the method of contrast variations in neutron scattering based on the use of various H$_2$O/²H$_2$O mixtures (Engelman and Moore 1975; Ibel and Stuhrmann 1975; Jacrot 1976; Ostanevich and Serdyuk 1982) was applied. The problem is that the scattering amplitude densities of H$_2$O and ²H$_2$O are much different in sign and value (-0.56 x 10^{10} cm^{-2} and $+6.39$ x 10^{10} cm^{-2}, respectively). The scattering amplitude densities of all biological macromolecules, including RNA and protein, are of intermediate values. Thus, contrast-match conditions can be achieved by placing the ribosome in various H$_2$O/²H$_2$O mixtures when one of the components (RNA or protein) becomes "invisible" for neutrons. Theoretical calculation of deuterium exchange and experiment show that the RNA component is contrast-matched in 70% ²H$_2$O (Serdyuk et al. 1980; Serdyuk 1979), whereas the protein is invisible in 40% to 42% ²H$_2$O (Jacrot 1976; Serdyuk 1979). Hence, based on the contrast variations by using different proportions of H$_2$O and ²H$_2$O in the solvent, the contribution of the protein component change of a ribonucleoprotein particle can be estimated, *independently of RNA component changes.*

Fig. 7. The dependence of the neutron scattering intensity normalized to the incoherent scattering of H_2O on the scattering vector μ in the Guinier coordinates [ln I $(\mu)_{normal}$ vs μ^2] for pretranslocation state ribosomes at three different contrasts. The wavelength $\lambda = 9.98$ Å, the spectral width $\triangle\lambda/\lambda = 8\%$, the sample-to-detector distances L = 5.5 m and 2.5 m, $t° = +4°C$. The following equation was used for the calculations of the normalized intensities:

$$I\,(\mu)_{normal} = \frac{[I(\mu)_{solution} - I(\mu)_{solvent}] \times (1\text{-}T) \times K}{4\ \pi\ell \times I_{H_2O}},$$

where ℓ is the cell length in cm, T is the transmission of solution, and K is the constant depending on the wavelength λ ($K = 5 \times 10^{-5}$ for $\lambda = 9.98$ Å). The temperature dependence of the incoherent scattering of H_2O was taken into account

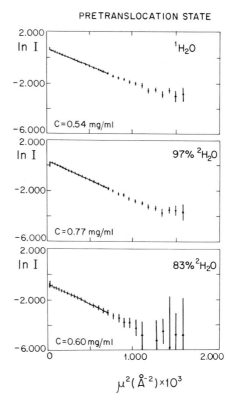

PRETRANSLOCATION STATE

Neutron scattering measurements were performed by the low angle camera D11 (Ibel 1976) at the high-flux reactor of the Institut Max von Laue-Paul Langevin, Grenoble. Figures 7 and 8 show the dependence of neutron scattering intensity I on scattering vector μ in the Guinier coordinates (ln I vs μ^2) for the pretranslocation and posttranslocation state ribosomes, respectively, at different contrast (H_2O, 97% ^2H_2O and 83% ^2H_2O). As seen from the figures, the Guinier plots for the translating ribosomes are straight lines within the scattering vector range of 0.006 Å^{-1} to 0.027 Å^{-1}. No rise of scattering intensities at the very small angles used ($\mu \sim 0.007$ Å) was observed either in H_2O or ^2H_2O, contrary to the previous reports on experiments with nontranslating 70S ribosomes prepared by standard methods (Stuhrmann et al. 1978). This indicates that an aggregated material, such as ribosome dimers, trimers, etc., was absent in noticeable amounts from the samples under investigation, as well as no aggregation was induced by incubation in ^2H_2O during measurements.

In order to check a possible effect of ^2H_2O on the compactness of the translating ribosomes, the X-ray radii of gyration of the ribosomes in the two functional states were also measured in the standard buffers with H_2O and 83% ^2H_2O. All electron radii of gyration of translating ribosomes were found to coincide in both H_2O and ^2H_2O, within the limits of experimental error (± 0.8 Å). Thus, ^2H_2O itself has no influence on the compactness of the ribosomes.

The dependence of the reduced neutron scattering intensity on the ^2H_2O fraction in $H_2O/^2H_2O$ mixture, in coordinates $\sqrt{I(0)_{normal}}$ / c·T vs % ^2H_2O, are straight lines

POST TRANSLOCATION STATE

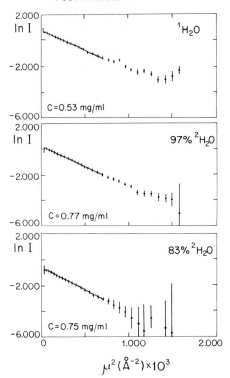

Fig. 8. The same as in Fig. 7 for posttranslocation state ribosomes at three different contrasts

parallel to each other; it is interesting to note that they intersect the abscissa at 59.5% and 58.5% ^2H_2O for the pretranslocation and posttranslocation state ribosomes, respectively (Fig. 9). This difference of the contrast-match points is consistent with the fact that the ribosome in the pretranslocation state contains one tRNA molecule more than the posttranslocation state ribosome.

The results of calculations of the neutron radii of gyration, including the data on the measurements of the ribosomes in 16% ^2H_2O, are summarized in Table 1. The values of the normalized intensity $I(0)_{normal}$ obtained by extrapolation of the Guinier plots to the zero scattering angle and divided by concentration of the ribosomal particles (reduced intensity) are also presented here. As seen from Table 1, the neutron radii of gyration of the ribosomal particles in the posttranslocation state are systematically greater than those of the pretranslocation state particles at all the contrasts used. At the same time, the value of the reduced normalized intensity of the neutron scattering (at zero scattering angle), which is proportional to a molecular mass, is found to be almost the same for the ribosomes in the two functional states (see Table 1). Hence, the small difference in R_g observed is not due to a contribution of an additional protein or RNA in one of the states. This observation leads to the conclusion that the difference in compactness of the ribosomes in the two functional states is a result of translocation.

Figure 10 demonstrates the dependence of the radii of gyration of the particles on the contrasts, in coordinates R_g^2 vs $1/\triangle\rho$, for the preparations of the pretranslocation and posttranslocation state ribosomes. It is seen that the difference in compactness in-

Table 1. Radii of gyration of the ribosomes in pretranslocation and posttranslocation states as measured by neutron scattering

Fraction of ^2H_2O in the $^1H_2O/^2H_2O$ mixture, vol. %	Radius of gyration, Å		Normalized scattering intensity extrapoled to zero scattering angle and divided by ribosome concentration $I(O)_{normal}$ / c	
	Pretranslocation	Posttranslocation	Pretranslocation	Posttranslocation
16	86.8±0.4	87.3±0.4	1.04±0.01	0.97±0.01
0	87.9±0.8	88.3±0.7	1.73±0.02	1.71±0.02
97	95.1±0.3	96.4±0.3	1.35±0.01	
83	98.4±0.8	101.0±0.9	0.49±0.01	0.50±0.01

Increase of protein contribution

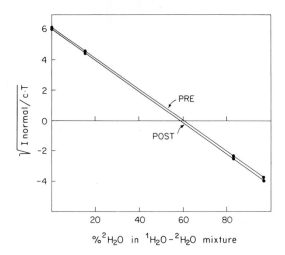

Fig. 9. The dependence of the reduced neutron scattering intensity on the H_2O fraction in the $H_2O/^2H_2O$ mixture, in coordinates

$$\sqrt{I(0)}_{normal} / c \cdot T \quad vs. \ \% \ ^2H_2O,$$

for pretranslocation and posttranslocation state ribosomes

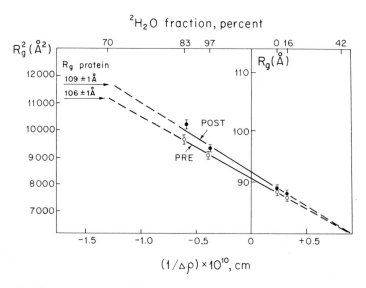

Fig. 10. The dependence of the radius of gyration of the pretranslocation (-○-) and posttranslocation (-●-) state ribosomes on the contrast in the Stuhrmann coordinates [R_g^2 vs $1/(\overline{\Delta\rho})$] (see Ibel and Stuhrmann 1975). The error values (±1.5 σ) are indicated by the *vertical* segments

creases as the scattering contribution of protein component becomes greater, attaining the maximum value of $\triangle R_g$ about 3 to 5 Å at the contrast where the contribution of the RNA component is negligible.

Thus, the experimental results reported here are the first to demonstrate that the translocation is accompanied by a relative spatial displacement of some parts of the ribosome, with the amplitude of several angstroms. The question is immediately raised which parts of the ribosome are involved in the movements, and whether a shift of one ribosomal subunit relative to the other is responsible for the difference observed in the compactness of the ribosomes of two functional states. This question is still open.

Acknowledgment. The preparation of translating ribosomes, their functional analyses, and the sedimentation experiments were made with participation of Vladimir Baranov, Institute of Protein Research, Pushchino. The electron microscopy study of the translating ribosomes was performed by Dr. Victor Vasiliev and Olga Selivanova, Institute of Protein Research, Pushchino. The neutron scattering experiments were conducted in collaboration with Dr. Roland May, Institut Laue-Lange-vin, Grenoble.

References

Baranov VI, Belitsina NV, Spirin AS (1979) The use of columns with matrix-bound polyuridylic acid for isolation of translating ribosomes. Methods Enzymol 59: 382-398

Baranov VI (1983) Preparation of translating ribosomes using columns with immobilized poly-uridylic acid, Bioorg Khim (USSR) 9: 1650-1657

Belitsina NV, Elizarov SM, Glukhova MA, Spirin AS, Butorin AS, Vasilenko SK (1975) Isolation of translating ribosomes with a resin-bound poly-U column, FEBS Lett 57: 262-266

Chuang DM, Simpson MV (1971) A translocation associated ribosomal conformational change detected by hydrogen exchange and sedimentation velocity. Proc Natl Acad Sci USA 68: 1474-1478

Conway TW, Lipmann F (1964) Characterization of a ribosome-linked guanosine triphosphatase in *E. coli* extracts. Proc Natl Acad Sci USA 52: 1462-1469

Engelmann DM, Moore PB (1975) Determination of quaternary structure by small angle neutron scattering. Annu Rev Biophys Bioeng 4: 219-241

Gupta SL, Waterson J, Sopori ML, Weissman SM, Lengyel P (1971) Movement of the ribosome along the messenger ribonucleic acid during protein synthesis. Biochemistry 10: 4410-4421

Ibel K, Stuhrmann HB (1975) Comparison of neutron and X-ray scattering of dilute myoglobin solution. J Mol Biol 93: 255-265

Ibel K (1976) The neutron small angle camera Dll at the high flux reactor, Grenoble. J Appl Cryst 9: 296-309

Jacrot B (1976) The study of the biological structures by neutron scattering from solution. Rep Prog Phys 39: 911-953

Lake JA (1976) Ribosome structure determined by electron microscopy of *E. coli* small subunits, large subunits and monomeric ribosomes. J Mol Biol 105: 111-130

Lipmann F (1969) Polypeptide chain elongation in protein biosynthesis. Science (Wash DC) 164: 1024-1031

Nishizuka Y, Lipmann F (1966) The interrelationship between guanosine triphosphatase and amino acid polymerization. Arch Biochem Biophys 116: 344-351

Ostanevich Yu M, Serdyuk IN (1982) Neutronographic studies of biological macromolecule structures, Usp Fiz Nauk 137: 85-116

Schreier MH, Noll H (1971) Conformational changes in ribosomes during protein synthesis, Proc Natl Acad Sci USA 68: 805-809

Serdyuk IN (1979) A method of joint use of electromagnetic and neutron scattering: a study of internal ribosomal structure. Methods Enzymol 59: 750-775

Serdyuk IN, Schpungin JL, Zaccai G (1980) Neutron scattering study of the 13S fragment of 16S RNA and its complex with ribosomal protein S4. J Mol Biol 137: 109-121

Spirin AS (1968) On the mechanism of the working of the ribosome. Hypothesis of locking-un-locking subparticles. Dokl Acad Nauk SSSR 179: 1467-1470

Spirin AS (1969) A model of the functioning ribosome: locking and unlocking of the ribosome sub-particles. Cold Spring Harbor Symp Quant Biol 34: 197-207

Stuhrmann HB, Koch MH, Parfait R, Haas J, Ibel K, Crichton RR (1978) Determination of the distribution of protein and nucleic acid in the 70S ribosomes of *Escherichia coli* and their 30S subunits by neutron scattering. J Mol Biol 119: 203-212

Vasiliev VD, Selivanova OM, Baranov VI, Spirin AS (1983) Structural study of translating 70S ribosomes from *E. coli* I. Electron microscopy. FEBS Lett 155: 167-172

Waterson J, Sopori ML, Gupta SL, Lengyel P (1972) Apparent changes in ribosome conformation during protein synthesis. Centrifugation at high speed distorts initiation, pretranslocation and posttranslocaction complexes to a different extent. Biochemistry 11: 1377-1382

Watson JD (1964) The synthesis of protein on ribosomes. Bull Soc Chim Biol 46: 1399-1425

Proof-Reading in Translation

The Effect of "In Vivo" Incorporation of α-Aminobutyric Acid into *E. coli* Proteins

T. Yamane[1], C.Y. Lai[2] and J.J. Hopfield[1, 3]

1 Introduction

There are only 20 primary amino acids specified in the genetic code, yet some 140 amino acids and amino acid derivatives are found in various proteins in different organisms (Uly and Wold 1977). Living organisms were not designed to use 20 primary amino acids; it evolved by random mutation. Since the rules of evolution remain the same now as always, i.e., nothing is immutable, it should be possible to force a cell to grow on an amino acid analog.

The present work was initiated with an herculian task of isolating an *E. coli* mutant with an ability to grow in the presence of a valine analog, α-aminobutyric acid (ABA); not an ABA-resistant mutant, but one which is able to replace partially, if not totally, valine by ABA in proteins. Reports in the literature claiming that the amino acid ABA, once called butyrine, is present in many proteins have appeared from time to time. These claims, however, have not been substantiated (Greenstein 1961). This amino acid occurs in tripeptide ophthalmic acid (γ-glutamyl-α-amino-n-butyrylglycine) in eye lens, which is identical in structure to glutathione except for the replacement of the SH group by CH_3. Its function is unknown (Metzler 1977). ABA is found in appreciable concentrations in normal blood and urine (Westfall et al. 1954a, 1954b; Gordon 1949) and at a higher concentrations in the urine of Fanconi syndrome patients (Dent 1946; Stanburg et al. 1966). The ratio ABA/leucine in plasma has been reported to increase in certain alcohol-related and unrelated liver diseases (Morgan et al. 1977).

We report here an observation made while studying the "biological fitness" in the presence of ABA of a genetically uncharacterized valine-isoleucine auxotroph of *E. coli*, strain AB3601, which was derived from K-12, and obtained from Coli Stock Culture Collection.

2 Recognition: Proofreading

A basic problem in specificity is the recognition by an enzyme of a substrate which is smaller than or isosteric with the biologically "correct" substrate. On the basis of

---

[1] Bell Laboratories, Murray Hill, New Jersey 07974
[2] Roche Institute of Molecular Biology, Nutley, New Jersey 07110
[3] California Institute of Technology, Pasadena, California 91125

Mechanisms of Protein Synthesis
ed. by Bermek
© Springer-Verlag Berlin Heidelberg 1984

Table 1. Activity of L-valine analogs with valyl-tRNA synthetase (ATP-PP$_i$ exchange reaction) (data from Bergmann et al. 1961)

Substrate	$K_m(M)$	$V_{max}(\%)$
DL-valine	1×10^{-4}	100
DL-threonine	1.2×10^{-2}	30
DL-α-aminobutyric acid (ABA)	3.7×10^{-3}	30
DL-threo-α-amino-β-chloro butyric acid (CBA)	3.3×10^{-4}	100
DL-α-aminoisobutyric acid (norvaline)		0

energetic considerations it has been calculated that in a Michaelis-Menten enzyme-substrate interaction with one recognition step, discrimination against similar isosteric compounds is no more than 1 in 20 (Pauling 1957). Isoleucyl-tRNA synthetase with a binding site for isoleucine cannot completely exclude the smaller valine which as a H atom in place of CH$_3$ group (Bergmann et al. 1961). The enzymatic discrimination between the two substrates is proportional to the differences in their potential binding energies ($\triangle G$) and can't be increased beyond this by any mechanism such as strain, induced fit, etc. The error fraction, f_0, is given by $f_0 = e^{-\triangle G/RT}$. One way of lowering the total f_0 is to have a succession of discriminating processes and holding the total f_0 to the biologically tolerable level. The kinetic proofreading theory provides an explanation for how this might be achieved with irreversible processes and a branched point on the pathway through which the undesired product can be channeled (Hopfield 1974; Ninio 1975). If a series of discriminatory processes is present, irreversible steps are needed to achieve a higher accuracy than individual steps. Nonspecific destructive reactions like apparently futile ATPase or GTPase are required for this purpose.

Table 1 shows the ATP-PP$_i$ exchange reaction of valine analogs with valyl-tRNA synthetase (E) (Bergmann et al. 1961). Threonine which differs from valine in the replacement of a CH$_3$ by an OH group has a 100-fold greater K$_m$ than valine and the maximum rate is 30% of that found with valine. β-chlorobutyric acid (CBA), in which the Cl atom has the same configuration as the OH group of threonine, is as active as valine (same V$_{max}$); although the K$_m$ is slightly higher, it can be transferred to tRNA^{Val} (Freundlich 1967). Substitution of one of the CH$_3$ groups of valine by H as in α-aminobutyric acid (ABA) results in an increase of the K$_m$ and a decrease in the V$_{max}$ and the complex E(ABA.AMP) fails to be transferred to tRNA^{Val} (Freundlich 1967). The intermediate complex E(ABA.tRNA^{Val}) is formed at the rate constant of 2.4 s^{-1} and hydrolyzed to E + ABA + tRNA^{Val} at an rate of 50 s^{-1} (Fersht and Dingwall 1979).

One point that should be stressed is the fact that most of tRNAs in *E. coli* cell is aminoacylated, bound to Tu.GTP factor (Furano 1975). The rate of incorporation of the amino acids from the Tu.GTP bound aminoacyl-tRNA into protein is faster than its enzyme-catalyzed deacylation (Fersht and Dingwall 1979). For instance, it is not possible to detect the misacylated Val-tRNA^{Ile} in the absence of Tu.GTP (Hopfield et al. 1976). For the reaction

$$E(Val\text{-}tRNA^{Ile}) \rightarrow E + Val + tRNA^{Ile}$$

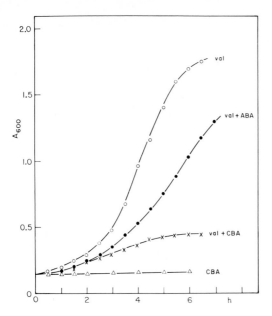

Fig. 1. Cell growth curve of *E. coli* AB3601, valine, and isoleucine auxotroph. Growth medium: SSA (g/l) K_2HPO_4, 10.5; KH_2PO_4, 4.3; Na citratel$_2$ H_2O, 0.47; ammonium sulfate, 1.0, $MgSO_4$ $_7H_2O$, 0.1; glucose, 0.2 %, thiamine, $\mu g/me$, supplemented with valine, 1 mM; isoleucine, 1 mM; plus ABA, 1 mM or CBA, 1 mM. -o- Control; -•- plus ABA; -X- plus CBA; $\triangle$ without valine plus CBA or ABA

the rate of hydrolysis is 10 s^{-1} in the absence of Tu-GTP and 0.01 s^{-1} in its presence (Fersht and Dingwall 1979). This turnover number for the hydrolysis is 100 times slower than the rate of turnover of aminoacyl-tRNA during protein synthesis, 1 s^{-1}. Thus, if there is an excess of Tu.GTP in the cell, the amino acid from misacylated tRNA released from the synthetase will be incorporated into proteins. It is rather ironic that the presence of Tu.GTP increases the misincorporation of wrong amino acids into proteins. However, the protein synthesis is faster in its presence; Tu.GTP is essential for the kinetic proofreading at the ribosomal site and it has a strong discriminatory power against the D-amino acids (Yamane et al. 1981).

Although the complex E(ABA.tRNA^{Val}) is hydrolyzed at a rate of 50 s^{-1}, the presence of Tu.GTP should lower this value by a factor of 10^2 to 10^3 to a value lower than 1 s^{-1} required for the incorporation into proteins in *E. coli*. How much ABA does get incorporated into *E. coli* proteins in vivo? This was one number we wanted to determine.

3 Effect of ABA on Cell Growth

Figure 1 shows the growth of *E. coli* AB3601 cells in the SSA medium supplemented with valine (1 mM), isoleucine (0.6 mM), plus ABA (1 mM) or CBA (1 mM). ABA or CBA alone cannot support the cell growth in the absence of valine. Although both ABA and CBA have an inhibitory effect on the growth, there is a major difference between these two valine analogs. This is not surprising. As discussed above, ABA and CBA show quite a different activity toward the valyl-tRNA synthetase. Moreover, ABA has no effect on derepression of the *ilv* enzymes in a mutant auxotroph for isoleucine, valine, and leucine when grown in limiting amounts of valine and isoleucine. In contrast, the addition of CBA to cells grown on limiting valine completely repressed the *ilv* enzymes (Freundlich 1967).

At the same concentration, CBA has a much stronger inhibitory effect than ABA and the cell growth eventually stops after barely doubling the cell density. This effect of CBA is very similar to the one observed with canavanine, an arginine analog, which causes cell growth inhibition and cell death due to a nondiscriminatory incorporation of this analog into the cell proteins (Schachtele and Rogers 1965; Schachtele et al. 1968).

In *E. coli* K-12, two isoenzymes, acetohydroxy acid synthase I and III (which are specified by the *ilv* B and *ilv* HI genes and catalyze the first step in valine and leucine biosynthesis from pyruvate and also the second step in isoleucine biosynthesis from threonine) are strongly inhibited by valine (Fig. 2). When this occurs, isoleucine biosynthesis is blocked and the cells are unable to grow. The present strain is not a leucine auxotroph and one would expect the intactness of the valine- sensitive *ilv* B and *ilv* HI genes required for the leucine biosynthesis. However, inhibition due to a high concentration of valine was not observed. This resembles an *E. coli* K-12 valine-resistant mutant described by Smith et al. (1979) which expressed *ilv* G gene which specifies the valine-insensitive acetohydroxy synthase II. Probably, *ilv* G gene is active in the present strain. Whether the *ilv* G gene product is ABA sensitive is not known.

In *E. coli*, the amino acids valine, isoleucine, and leucine are taken up by cells through a common transport system (Oxender 1972). The growth inhibition observed is due, partly, to the reduced transport rate of valine caused by the competition of ABA for the same binding protein. How strongly it competes with valine was not determined. Isoleucine competes rather efficiently against valine in the transport and if an excess of isoleucine over valine is required for growth, usually a dipeptide, glycylvaline, is added instead of valine. In the present strain, isoleucine concentration higher than those of valine caused growth inhibition, as expected.

Besides inhibiting the valine transport, is ABA acting at the gene level? It is possible that ABA can interact with *ilv* or orther aporepressor(s) which may act as surrogate repressor for other operons. In fact, several mutants resistant to threonine and leucine analogs (DL-α-amino-β-hydroxy-valeric, 5', 5', 5'-trifluoro-DL-leucine and 4-azaleucine) confer constitutive expression of the *thr, leu,* and *ilv* operons, i.e., derepressed levels of the biosynthetic enzymes for the branched-chain amino acids. This indicates the existence of an interconnected regulatory network which could function to control several amino acid biosynthetic(s) systems in a coordinated manner (Johnson and Somerville 1983). Similarly, in the presence of tryptophan, cells with an elevated level of Trp aporepressor protein have a complex nutritional requirement (TROP syndrome) that is overcome by the addition of threonine, phenylalanine, leucine, serine, valine, and tyrosine. Trp aporepressor, when overproduced, can act as a repressor for the operons of these amino acids (Johnson and Somerville 1983).

As discussed above, the present strain AB3601 is, in a sense, a valine-resistant mutant and the valine-insensitive acetohydroxy synthase II is probably being produced. There is also a remote possibility that although the gene *ilv* G can't be repressed by valine, it is sensitive to ABA. If so, the growth inhibition observed is due partly to the repression of ilv G gene and the consequent halt of leucine biosynthesis. In fact, the inhibition caused by ABA can partly be reversed by the addition of leucine to the growth medium.

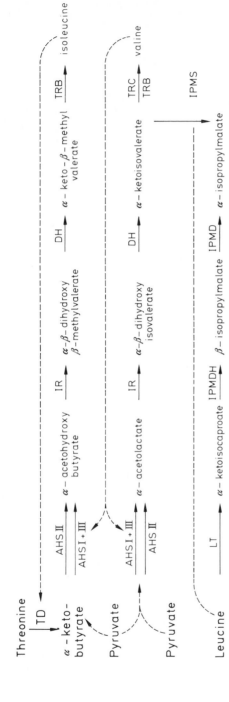

Fig. 2. Biosynthesis of isoleucine, valine, and leucine. The enzymes catalyzing the indicated steps are abbreviated and the corresponding structural genes are indicated in parentheses as follows: TD (*ilv* A), end product inhibited threonine deaminase; AHS I (*ilv* B) and AHS III (*ilv* HI), end product inhibited acetohydroxy-synthases (acetolactate synthase); AHS II (*ilv* G), end product-noninhibited acetohydroxy acid synthase; IR (*ilv* C), acetohydroxy acid isomeroreductase (acetolactate isomeroreductase); DH (*ilv* D), dihydroxy acid dehydrase; TRB (*ilv* E), transaminase B; TRC, transaminase C; IPMS, end product inhibited α-isopropylmalate synthase; IPMD, α-isopropylmalate dehydratase; IPMDH, isopropylmalate dehydrogenase; LT, leucine transaminase. The enzymes essential for the biosynthesis of valine, isoleucine, and leucine are encoded in the *ilv* and *leu* operons. These operons are thought to be controlled by attenuation mechanism similar to that established for *trp*, *his*, and *phe* operons (Johnson and Somerville, 1983; Artz and Proach, 1975; DiNocera et al., 1978; Zurawski et al., 1978)

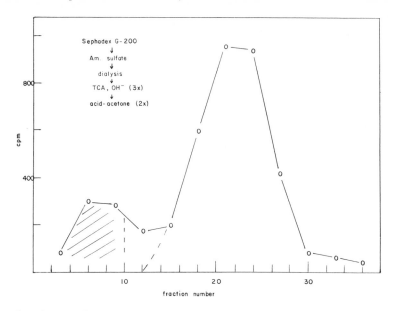

Fig. 3. Isolation of total protein for amino acid analysis. Cells were grown as described in the legend to Fig. 1., harvested at the middle log phase, pulverized in liquid nitrogen, and centrifuged; the supernatant made 4.5 M in urea and passed through Sephadex G-200 in order to eliminate free amino acids. The protein fraction was precipitated with ammonium sulfate (0.6 g/ml) and dialyzed against water for 24 h with frequent change of water. This process was followed by precipitation with TCA and the precipitate was dissolved in 0.1 M NaOH. This step was repeated three times and the amount of radioactivity remained constant, indicating that the protein sample was not contaminated with free amino acids. The final step was the acid-acetone treatment which was repeated twice

4 Amino Acids Analysis

In order to determine the amount of ABA incorporated into proteins, *E. coli* cells were grown in the presence of [^{14}C]-aminobutyric acid, total protein isolated (Fig. 3) and amino acid analysis was performed on a Durham DC4A HPLC column under conditions specified in the legend to Fig. 4 and amino acids detected by the fluorescamine method (Stein et al. 1973). The analysis concentrated on four amino acids: glycine, alanine, valine, and ABA. Table 2 shows the content of these amino acids in proteins isolated from cells grown in the presence and absence of ABA. No difference in the content of these amino acids was observed and the amount of ABA incorporated into proteins was 1.6 ABA per 1000 valine when cells were grown in the presence of 1 mM ABA and 1 mM valine. In *E. coli* valine constitutes about 9% of proteins (Raunio and Rosenqvist 1970). Therefore, the amount of ABA incorporated constitutes about 0.015% of the total protein, a negligible concentration due to an efficient proofreading process.

Given the amount of ABA incorporated into proteins, it is rather difficult to understand its effect on cell growth. At a concentration of 1 mM in both valine and ABA, the cell duplication rate was reduced by a factor of 2. The amount of energy spent for proofreading is not sufficient to have any detrimental effect on growth.

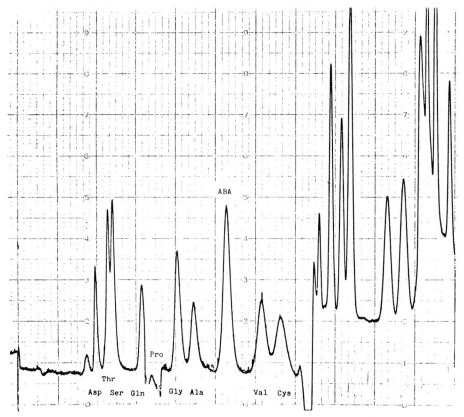

Fig. 4 a-b. Amino acid analysis. Amino acid analysis was carried out on a Durham DC-4A HPLC column and amino acids were detected by fluorescamine method. The quantitative analysis concentrated on four amino acids: glycine, alanine, valine, and ABA. In order to minimize the overlapping of the peaks in the ABA region, cysteine was oxidized with performic acid to cysteic acid which elutes before aspartate, and chromotography was carried out with the first elution buffer acidified with 0.4 ml 12N HCl per liter.

a Standard amino acids; **b** hydrolysate of the total protein from the cells grown in the presence of ABA

Table 2. Amino acid content[a] of proteins isolated from *E. coli* treated and untreated with ABA

	-ABA	+ABA
Gly	0.45	0.45
Ala	0.52	0.51
Val	0.36	0.37
ABA	Not detectable	0.0006

[a] μmol per mg protein

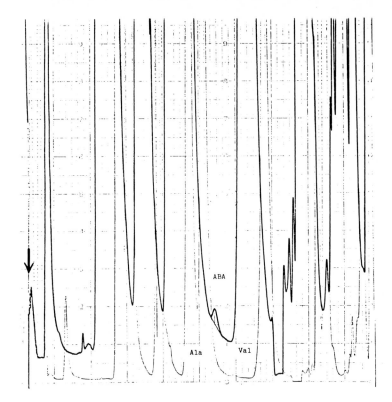

Fig. 4b

5 Effect of ABA on β-Galactosidase Activity

Are the proteins synthesized in the presence of ABA functional? A priori, since the amount of ABA incorporated is negligible, it is obvious to expect the proteins to be equally functional as those synthesized in the absence of this valine analog. To answer this question, *E. coli* cell culture at the cell density of 0.2 (A_{600}) was divided into two fractions and to one ABA was added at a concentration of 1 mM and the β-galactosidase activity was determined as the cell density increased. In Fig. 5 the activity is plotted in Units/mg protein and it shows that the enzyme activity in ABA-treated cells is 30% lower than that of the control. This is rather puzzling considering the fact that the amount of ABA replacing valine in the cell is known to be negligible. The low enzyme activity observed is not unique to β-galactosidase. RNA polymerase also showed a lower activity compared to the control (data not shown).

6 Polyacrylamide Gel Electrophoresis of Proteins

The low β-galactosidase activity observed in the cells grown in the presence of ABA led us to ask whether abnormal proteins were being synthesized. Organisms live utterly at the mercy of the environment and have to adjust to any change in nutritional or physical

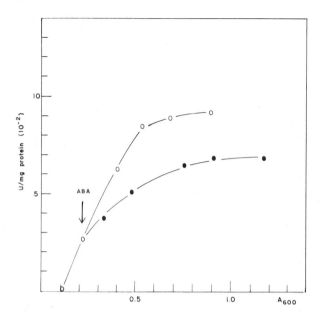

Fig. 5. β-galactosidase activity. The procedure described in Experiments in Molecular Genetics by J.H. Miller (Cold Spring Harbor Laboratory, 1972, p. 352) was followed for the assay of enzyme activity. Activity is expressed μmol of product formed per mg protein per min. -○- control; -●- plus ABA

conditions. Any stress imposed on cell cultures or organisms, including treatment with amino acid analogs (Johnston et al. 1980), seems to induce a series of characteristic stress-proteins or heat-shock-proteins in the 60 000-70 000 and 80 000-90 000 mol. wt. range (Schlesinger et al. 1982).

Figure 6 shows the polyacrylamide gel electrophoresis of protein samples from ABA-treated and untreated cells. There are missing or fainter bands in samples from ABA-treated cells compared to the control samples, namely, 18600, 29400, and 38900 mol. wt. proteins. The β-galactosidase monomer is 135000 mol. wt. protein and it can't be identified in this gel. The identity of these missing bands are not known at this moment.

7 Proteolysis

The obvious explanation for the missing bands would be to assume a preferential proteolysis of these proteins. It is a well-established fact that when exponentially growing bacteria are deprived of nitrogen or carbon source, or a required amino acid, the rate of catabolism of cellular protein increases several fold (Pine 1967). Similarly, *E. coli* proteins synthesized in the presence of amino acid analogs are degraded more rapidly than those containing normal amino acids (Goldberg et al. 1975).

In order to test whether the incorporation of ABA at low level increase the cell proteolysis, cells were grown first in the SSA medium containing normal amino acids. Cells were collected during the exponential growth and resuspended in original growth medium or in medium containing valine (1 m*M*) plus ABA (1 m*M*) or CBA (1m*M*) in the presence of [³H]Leu for two generations, harvested, washed, and resuspended in

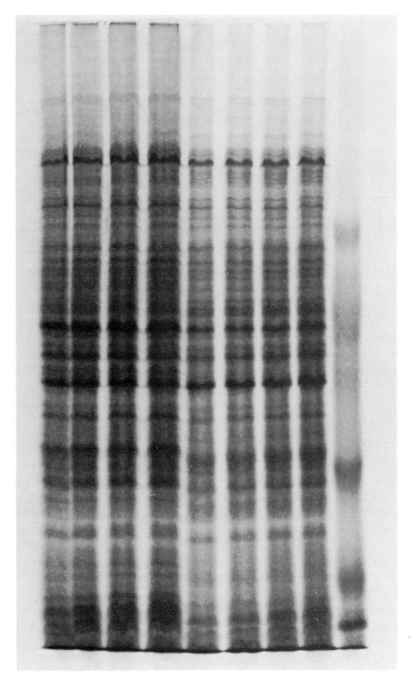

Fig. 6. Polyacrylamide gel electrophoresis of proteins from ABA-treated and untreated cells. Electrophoresis was performed on 16×25 cm slab gels using the discontinuous buffer system of Laemmli (1970). Separating were 15% polyacrylamide with 0.375 M Tris-HCl and 0.1% SDS, pH 8.8; stacking gels were 5% polyacrylamide with 0.125 M Tris-HCl, 0.1% SDS, pH 6.8. The electrophoretic buffer was 0.025 M Tris-HCl, 0.192 M glycine, 0.1% SDS, pH 8.3. Electrophoresis was run at room temperature at a constant voltage of 130 V and approx. 15 h. From left, 1-4 untreated; 5-8, ABA-treated; 9, standards

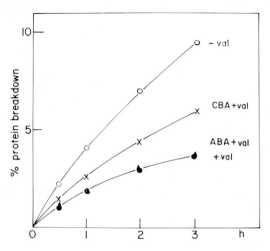

Fig. 7. Degradation of proteins from cells grown in the presence of valine analogs. *E. coli* AB3601 was grown in SSA medium supplemented with required amino acids (val 1 mM + isoleucine 0.6 mM), then exposed to [^{3}H]Leu in the presence of ABA (1 mM) or CBA (1 mM). The cells were collected during exponential growth, washed, and resuspended in the original growth medium supplemented with cold leucine. Catabolism of proteins containing valine (-●-); ABA (-▲-); CBA (-X-); cells resuspended in the medium without valine (-○-)

medium containing only valine plus cold leucine and the protein breakdown was checked. The amount of [^{3}H]Leu in TCA soluble supernatant was taken as the amount of proteolysis. As shown in Fig. 7, the proteolysis of cells grown in the presence of ABA is indistinguishable from that of control, indicating the incorporation of negligible amount of ABA is not sufficient to trigger an increase in the proteolysis. A much higher rate of catabolism is observed in cells grown in the presence of CBA, due to a high incorporation of this valine analog into proteins.

8 Conclusion

The growth inhibition of prokaryotes due to ABA is a well-known fact (Gladstone 1939; Rowley 1953; Wiame and Storck 1953; Friedman 1955; Freundlich and Clarke 1968). ABA is a potent inhibitor of (a) acetohydroxy acid synthase I and III, the initial enzyme common to isoleucine, leucine, and valine biosynthesis (Freundlich and Clarke 1968), (b) valine transport through the cell membrane (Oxender 1972), and (c) activation and the transfer of valine to tRNA^{Val} (Freundlich 1967; Freundlich and Clarke 1968). It has no effect on derepression of the *ilv* enzymes (Freundliche 1967). However, these facts are not sufficient to explain the present observations.

The present strain AB3601 is valine-resistant, i.e., no growth inhibition is observed at high valine concentration. This implies that since the cell is not a leucine auxotroph, *ilv* G gene is active and synthesizes valine-resistant acetohydroxy acid synthase II which is essential for leucine biosynthesis. There is a remote possibility that although *ilv* G gene product is valine-resistant, it is partly ABA-sensitive. In that case, leucine should reverse the inhibition. In fact, leucine is somewhat effective in reducing the growth inhibition caused by the analog. On the other hand, it has been shown that the keto analog of ABA, α-keto-butyrate inhibits the activity of α-isopropylmalate synthase, the first enzyme specific for leucine biosynthesis (Kohlhaw and Umbarger 1965). The conversion of in vivo of ABA to α-ketobutyrate (Umbarger, McElroy and Glass 1956) may account for some of the effects of ABA on growth.

Two aspects of this investigation remain puzzling: the absence of certain protein bands on the gel electrophoresis and the lower activity of certain or, possibly, all enzymes. The amount of ABA incorporated into proteins replacing valine is about 0.02%, not sufficient to cause a massive proteolysis due to nonfunctional proteins. Whether ABA causes the stringent response and accumulation of ppGpp in the cell is not known. However, the difference in the protein gel electrophoretic pattern observed suggest a shift in the levels of genes that are transcribed. The synthesis of stress proteins was not observed.

Cells grown in the presence of ABA look quite normal under the microscope although it takes longer than the control cells to form observable colonies on agar plates containing normal growth medium. Each cell culture has its characteristic odor depending on the growth medium used. The addition of ABA produces a quite distinct odor indicating a change in the metabolism of the cell.

At this moment we have few tentative answers to the observations reported here. However, it is clear that the observed physiological and biochemical effects of the analog are multivaried and, definitely, cannot be attributed to effects at a single site.

Acknowledgement. We thank M. Freundlich and P. Margolin for helpful discussions and B. Wyluda for technical assistance. This work was supported in part by NSF Grant DMR-8107494 to J.J.H.

References

Artz SW and Broach JR (1975) Histidine regulation in S. *typhimurium:* an activator-attenuator model of gene regulation. Proc Natl Acad Sci USA 72: 3453-3457

Bergmann FH, Berg P and Dieckmann M (1961) The enzymic synthesis of amino acyl derivatives of ribonucleic acid. J Biol Chem 236: 1735-1740

Cohen GN (1958) Synthese de proteines "anormales" chez *E. coli* K 12 cultive en presence de L-valine. Ann Inst Pasteur (Paris) 94: 15-30

Dent CE (1946) Partition chromatography on paper applied to the investigation of amino acids in peptides in natural and pathological urines. Biochem J 40: xliv-xlv

DiNocera PP, Blasi F, DiLaure R, Frunzio R and Bruni CB (1978) Nucleotide sequence of the attenuator region of the histidine operon of *E. coli* k-12. Proc Natl Acad Sci USA 75: 4276-4280

Fersht AR and Dingwall C (1979) Establishing the misacylation/deacylation of the tRNA pathway for the editing mechanism of prokaryotic and eukaryotic valyl-tRNA synthetases. Biochemistry 18: 1238-1244

Freundlich M (1967) Valyl-tRNA: role in repression of the isoleucine-valine enzymes in *E. coli.* Science (Wash DC) 157: 823-825

Furano AV (1975) Content of elongation factor Tu in *E. coli.* Proc Natl Acad Sci USA 72: 4780-4784

Goldberg AL, Olden K and Prouty WF (1975) Studies of the mechanisms and selectivity of protein degradation in *E. coli.* In: Schimke RT and Katunuma N (eds) Intracellular protein turnover. Academic New York, pp 17-55

Gordon AH (1949) An investigation of the intracellular fluid of calf embryo muscle. Biochem J 45: 99-105

Greenstein JP and Winitz M (1961) Chemistry of the amino acids, vol I. Wiley, New York, pp 91-119

Hopfield JJ (1974) Kinetic proofreading: a new mechanism for reducing errors in biosynthetic processes requiring high specificity. Proc Natl Acad Sci USA 71: 4135-4139

Hopfield JJ and Yamane T (1980) The fidelity of protein synthesis. In: Chambliss G, Craven GR, Davies J, Davis K, Kahan L and Nomura M (eds) Ribosomes: structure, function and genetics. University Park Press, Baltimore, USA

Johnson DI and Somerville RL (1983) Evidence that repression mechanisms can exert control over the thr, leu and ilv operons of *E. coli* k-12. J Bacteriol 155: 49-55

Johnston D, Oppermann H, Jackson J and Levinson W (1980) Induction of four protein in chick embryo cells by sodium arsenite. J Biol Chem 255: 6975-6980

Lee F and Yanofsky C (1977) Transcription termination at the trp operon attenuators of *E. coli* and *S. typhimurium:* RNA secondary structure and regulation of termination. Proc Natl Acad Sci USA 74: 4365-4369

Meister A (1965) Biochemistry of the amino acids, 2nd edn, vol I. Academic, New York

Metzler DE (1977) Biochemistry. Academic, New York, p 421

Morgan MY, Milsom JP and Sherlock S (1977) Ratio of plasma α-amino-n-butyric acid to leucine as an empirical marker of alcoholism: diagnostic value. Science (Wash DC) 197: 1183-1185

Ninio J (1975) Kinetic amplification of enzyme discrimination. Biochimie (Paris) 57: 587-595

Ninio J and Chapeville F (1980) Recognition: the kinetic concepts. In: Chapeville F and Haenni AL (eds) Molecular biology biochemistry and biophysics, vol 32: chemical recognition in biology. Springer, Berlin Heidelberg New York, pp 78-85

Oxender DL (1972) Membrane transport. In: Annu Rev Biochem 41: 777-814

Pauling L (1957) The probability of errors in the process of synthesis of protein molecules. In: Arbeiten aus dem Gebiet der Naturstoffchemie (Festschrift Arthur Stoll). Birkhauser AG, Basel, pp 597-602

Pine MJ (1967) Response of intracellular proteolysis to alteration of bacterial protein and the implications in metabolic regulation. J Bacteriol 93: 1527-1533

Raunio R and Rosenqvist H (1970) Amino acid pool of *E. coli* during the different phases of growth. Acta Chem Scand 24: 2737-2744

Rudman D and Meister A (1953) Transamination in *E. coli.* J Biol Chem 205: 591-604

Schachtele C and Rogers P (1965) Canavanine death in *E. coli.* J Mol Biol 14: 474-489

Schachtele C, Anderson DL and Rogers P (1968) Mechanism of canavanine death in *E. coli.* J Mol Biol 33: 861-872

Schlesinger MJ, Aliperti G and Kelley PM (1982) The response of cells to heat shock. Trends Biochem Sci 7: 222-225

Smith JM, Smith FJ and Umbarger HE (1979) Mutations affecting the formation of acetohydroxy acid synthase II in *E. coli* K-12. Mol Gen Genet 169: 299-314

Stanburg JB, Wyngaarden JB and Fredrickson DS (1966) The metabolic basis of inherited disease, 2nd ed. McGraw-Hill, New York

Stein S, Böhlen P, Stone J, Dairman W and Udenfriend S (1973) Amino acid analysis with fluorescamine at the picomole level. Arch Biochem Biophys 155: 202-212

Uly R and Wold F (1977) Posttranslational covalent modification of proteins. Science (Wash DC) 198: 890-896

Westfall BB, Peppers EV and Earle WR (1954a) The free and combined amino acids in the protein-free ultrafiltrate from whole chick-embryo extracts. J Natl Cancer Inst 15: 433-438

Westfall BB, Peppers EV, Sanford KK and Earle WR (1954b) The amino acid content of the ultrafiltrate from horse serum. J Natl Cancer Inst 15: 27-35

Wiame JM and Storck R (1953) Metabolisme de l'acide glutamique chez *B. subtilis*. Biochim Biophys Acta 10: 268-279

Yamane T, Miller DL and Hopfield JJ (1981) Discrimination between D and L-tyrosyl transfer ribonucleic acids in peptide chain elongation. Biochemistry 20: 7059-7064

Zurawski G, Brown K, Kïllingly D and Yanofsky C (1978) Nucleotide sequence of the leader region of phenylalanine operon of *E. coli.* Proc Natl Acad Sci USA 75: 4271-4275

Zurawski G, Elseviers D, Stauffer V and Yanofsky C (1978) Translational control of transcription at the attenuator of the *E. coli* tryptophan operon. Proc Natl Acad Sci USA 75: 5988-5992

Regulation of Eukaryotic Chain Initiation

A Direct Correlation Between the Affinity of a Given mRNA for Eukaryotic Initiation Factor 2 and its Ability to Compete in Translation

R. KAEMPFER, H. ROSEN, and G. DI SEGNI[1]

1 Introduction

Eukaryotic gene expression is often regulated by the selective translation of specific mRNA templates over other ones. Examples of this type of control, involving competition between mRNA species, are encountered in cellular differentiation and virus infection. Messenger RNA competition is thought to occur mainly at initiation of translation, during the recognition of mRNA and its binding to ribosomes.

Among the initiation factors for eukaryotic translation, eukaryotic initiation factor 2 (eIF-2) stands out by its importance to translational control. The activity of eIF-2 is regulated in response to a large variety of biological stress conditions, such as heme deprivation (Kaempfer 1974; Clemens et al. 1975), the presence of double-stranded RNA (dsRNA) (Kaempfer 1974; Clemens et al. 1975), or after treatment with interferon (Kaempfer et al. 1979b). Here, evidence will be reviewed in support of the concept that eIF-2 interacts directly with mRNA during protein synthesis, recognizing in mRNA molecules the sequence and conformation that constitute the ribosome binding site, and that the ability of a given mRNA species to compete in translation is correlated directly with its affinity for eIF-2. These properties impart on eIF-2 an essential function in differential gene expression at the level of translation.

2 Results

2.1 The mRNA-Binding Activity of eIF-2

Binding of mRNA to the 40 S ribosomal subunit cannot take place unless methionyl-tRNA$_f$ (Met-tRNA$_f$) is first bound (Schreier and Staehelin 1973; Darnbrough et al. 1973). This means that the recognition and binding of Met-tRNA$_f$ are an integral part of the mRNA binding process. Met-tRNA$_f$ is bound with absolute specificity by eIF-2 (Chen et al. 1972). This binding requires GTP and leads to formation of a ternary complex, eIF-2/Met-tRNA$_f$/GTP, that subsequently binds to the 40 S ribosomal subunit (e.g., Levin et al. 1973). The unique property of providing Met-tRNA$_f$ already imparts on eIF-2 a crucial role in the binding of mRNA. While additional initiation factors

[1]   Department of Molecular Virology, The Hebrew University-Hadassah Medical School, 91 010 Jerusalem, Israel

Mechanisms of Protein Synthesis
ed. by Bermek
© Springer-Verlag Berlin Heidelberg 1984

participate in stable binding of mRNA (e.g., Trachsel et al. 1977; Grifo et al. 1982), none can act in the absence of eIF-2.

In addition to binding Met-tRNA$_f$, eIF-2 itself can bind to mRNA (Kaempfer 1974; Barrieux and Rosenfeld 1977, 1978; Kaempfer et al. 1978a; Kaempfer et al. 1979a; Chaudhuri et al. 1981). This binding is specific in that all mRNA species tested possess an effective binding site for eIF-2, including mRNA species lacking the 5'-terminal cap or 3'-terminal poly(A) moieties (Kaempfer et al. 1978a), while RNA species not serving as mRNA, such as tRNA (Barrieux and Rosenfeld 1977, 1978; Kaempfer et al. 1978a, 1979a; Rosen and Kaempfer 1979), ribosomal RNA (Barrieux and Rosenfeld 1977; Kaempfer et al. 1981), or negative-strand viral RNA (Kaempfer et al. 1978a) bind far more weakly.

The fact that eIF-2 binds to RNA in general, even though it prefers mRNA, initially raised some question as to the specificity of its interaction with mRNA (Hershey 1982). However, a common property of proteins that recognize specific sites in nucleic acids, such as the *lac* repressor, is their tendency to bind with low affinity to nonspecific sequences (Lin and Riggs 1975; von Hippel et al. 1974). In the case of eIF-2, specificity of the interaction with mRNA is now supported by both structural and functional evidence (see below).

The mRNA-binding property is a function of eIF-2 itself. Thus, eIF-2 is preferentially retained on mRNA-cellulose columns and upon elution is active in translation and in GTP-dependent binding of Met-tRNA$_f$ (Kaempfer et al. 1978a; Kaempfer 1979). Binding of Met-tRNA$_f$ is inhibited competitively by mRNA (Kaempfer et al. 1978b; Barrieux and Rosenfeld 1978; Rosen et al. 1981a; Chaudhuri et al. 1981). Most convincingly, binding of mRNA to purified eIF-2 preparations can be inhibited completely by competing amounts of Met-tRNA$_f$, provided GTP is present, but not by uncharged tRNA (Fig. 1A). Partly purified eIF-4B is also able to retain globin mRNA on filters, but this binding, by contrast, is not sensitive to competition by Met-tRNA$_f$ (Fig. 1B).

Thus, mRNA and Met-tRNA$_f$ are mutually exclusive in their binding to eIF-2, suggesting that during initiation of translation, the interaction between a molecule of mRNA and eIF-2 on the 40 S ribosomal subunit could displace the previously bound Met-tRNA$_f$ from this factor. Rosen and Kaempfer (1979) have proposed that during initiation, three processes may occur in one step: binding of mRNA to eIF-2, displacement of Met-tRNA$_f$ from eIF-2, and base-pairing between mRNA and Met-tRNA$_f$.

2.2 The Site in mRNA Recognized by eIF-2

Globin mRNA molecules lacking the 3'-terminal poly(A) tail or an additional 90 nucleotides from the 3'-untranslated region bind to eIF-2 as tightly as native globin mRNA, with an apparent dissociation constant of 5×10^{-9} M at 150 mM KCl (Kaempfer et al. 1979a). On the other hand, cap analogs inhibit binding of both mRNA and Met-tRNA$_f$ to eIF-2 (Kaempfer et al. 1978b). Although this could suggest that the cap in mRNA interacts with eIF-2, the genomic RNA species from mengovirus or satellite tobacco necrosis virus (STNV) bind extremely well to eIF-2, in fact even better than globin mRNA, yet they do not carry a cap structure (Kaempfer et al. 1978b, 1981; Rosen et al. 1982). This and the observation that eIF-2 prefers native globin mRNA by five orders of magnitude over cap analogs led to the suggestion that binding of eIF-2 to mRNA occurs primarily at an internal sequence (Kaempfer et al. 1978b).

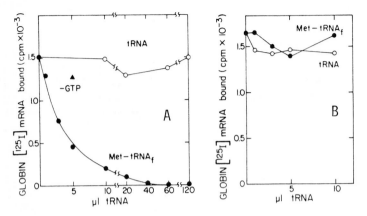

Fig. 1 A-B. Mutually exclusive binding of mRNA and Met-tRNA_f to eIF-2. Binding of ^{125}I-labeled globin mRNA (input, 2700 cpm) was studied in conditions for ternary complex formation in the presence of GTP, highly purified eIF-2 **A** or partially purified eIF-4B **B**, and the indicated volumes of mouse liver tRNA (0.1 mg/ml) charged in the absence (tRNA) or presence of L-methionine (Met-tRNA_f). Controls show that Met-tRNA_f is not hydrolyzed during incubation as in **B**. From Rosen and Kaempfer (1979)

2.2.1 A Relationship Between Binding Site Sequences for Ribosomes and for eIF-2

Mengovirus RNA has a length of about 7 500 nucleotides and contains a poly(C) tract located several hundred nucleotides from the 5' end. In 40 S or 80 S initiation complexes on this RNA, formed in L cell or ascites cell lysates, four specific T1 oligonucleotides, 15-28 bases in length, are protected by ribosomes against nuclease attack (Perez-Bercoff and Kaempfer 1982). These map into at least two widely separated domains at internal sites located downstream from the poly(C) tract. When intact mengovirus RNA is offered to eIF-2 in the absence of any other component, and the sequences protected by the initiation factor are isolated, three specific oligonucleotides are recovered out of this very large sequence, and these are identical with three of the four protected in initiation complexes (Perez-Bercoff and Kaempfer 1982). Thus, in mengovirus RNA, the binding sites for ribosomes on the one hand, and for eIF-2 alone on the other, are virtually identical. This finding suggests that binding of ribosomes to mengovirus RNA may be guided to an important extent by eIF-2.

A more detailed study was made of the binding site for eIF-2 in STNV RNA (Kaempfer et al. 1981). This RNA, 1 239 nucleotides long, has an unmodified 5' end that can be labeled in vitro with polynucleotide kinase. RNA isolated from STNV virions migrates, after 5' end labeling, as a heterogeneous collection of fragments in gels, with only a minor amount of label in fully intact viral RNA. Such preparations contain some 35 different 5' end sequences, as judged by fingerprinting with pancreatic or T1 ribonuclease (Kaempfer et al. 1981), attesting to the presence of many fragments originating from internal regions of the viral RNA molecule. When this RNA is offered to eIF-2 and RNA bound by eIF-2 is isolated and fingerprinted, one major spot is recovered, migrating precisely as the 5' end of intact STNV RNA (Kaempfer et al. 1981). Hence, eIF-2 binds selectively to STNV RNA fragments possessing the 5' end of the intact RNA.

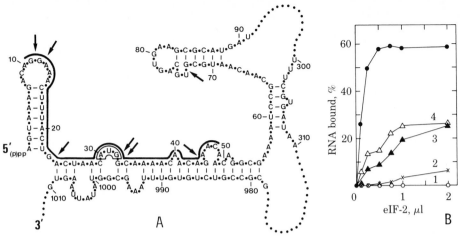

Fig. 2 A, B. Secondary structure model for the 5' end of STNV RNA **A** and eIF-2 saturation binding curves for intact STNV RNA and individual 5'-terminal fragments **B**. The model in **A** depicts stable secondary interactions. *Arrows,* prominent sites of RNase T1 cleavage. *Line,* nucleotides protected by 40 S ribosomal subunits. See text. In **B**, binding of intact STNV RNA (*top curve*) and of isolated 5'-terminal fragments with a length of 32 (*curve 1*), 44 (*curve 2*), 73 (*curve 3*), and 115 (*curve 4*) nucleotides, respectively (see **A**), was determined in the presence of the indicated amounts of eIF-2 (0.15 μg/μl). From Kaempfer et al. (1981)

To map the eIF-2 binding site more exactly, intact 5' end-labeled STNV RNA was isolated and digested partially with RNase T1, to generate a nested set of labeled RNA fragments, all containing the 5' end of intact RNA and extending to various points within the molecule. Fragments of discrete size were isolated and their ability to bind to eIF-2 was studied. Figure 2A depicts the 5'-terminal sequence of STNV RNA, including the unique AUG translation initiation codon located at positions 30-32. Arrows denote G residues sensitive to RNase T1 attack. eIF-2 does not bind the 32-nucleotide 5'-terminal fragment ending with the AUG codon, or shorter ones, but it does bind to the 44-nucleotide fragment or larger ones (Fig. 2B). This binding is with the same specificity as to intact viral RNA (Kaempfer et al. 1981), even though, as can be seen from Fig. 2B, the affinity of eIF-2 for short STNV RNA fragments increases with their nucleotide length. This result places the 3'-proximal boundary of the eIF-2 binding site at or near nucleotide 44. Indeed, binding of eIF-2 to intact STNV RNA greatly increases the sensitivity of this RNA to cleavage by RNase T1 at nucleotide 44, and by nuclease P1 near position 60 (Kaempfer et al. 1981). Apparently, binding of the initiation factor molecule to STNV RNA induces a considerable conformational change in the region just downstream from the binding site.

On the 5'-terminal side, eIF-2 shields positions 11, 12, 23, 32, and 33 against digestion (Fig. 2A), placing the boundary at or before position 10. Since the G residues at positions 2 and 7 are hydrogen-bonded and thus resistant to nuclease attack (Leung et al. 1979), it is not certain if the eIF-2 binding site extends to the physical 5' end. The striking aspect of the eIF-2 binding site is, however, that it overlaps virtually completely with the binding site for 40 S ribosomal subunits (Browning et al. 1980), depicted by the line in Fig. 1. Thus, eIF-2 by itself recognizes virtually the same sequence in STNV RNA that is bound by 40 S ribosomal subunits carrying eIF-2, Met-tRNA$_f$, and all other

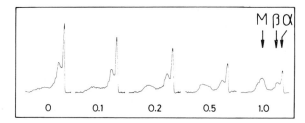

Fig. 3. Cellulose acetate electrophoretic analysis of products synthesized during simultaneous translation of globin mRNA and mengovirus RNA. Translation mixtures contained 1.1 μg of globin mRNA and the indicated μg amounts of mengovirus RNA. Densitometer scans of the autoradiogram of ^{35}S-labeled products are shown. α, α-globin; β, β-globin; M, mengovirus RNA-directed products of translation. From Rosen et al. (1982)

components needed for initiation of translation. These results extend the findings with mengovirus RNA (Perez-Bercoff and Kaempfer 1982) and point to a critical role for eIF-2 in the recognition of mRNA by ribosomes.

2.3 Translational Control by mRNA Competition for eIF-2

To study the functional implications of the interaction between mRNA and eIF-2, we analyzed translational competition, choosing to work with the mRNA-dependent reticulocyte lysate because it allows the precise quantitation of each mRNA species during translation and is capable of repeated initiation with an efficiency approaching that of the intact cell (Pelham and Jackson 1976; Jackson 1982).

2.3.1 Competition Between Mengovirus RNA and Globin mRNA for eIF-2

Evidence that individual mRNA species possess distinctly different affinities for eIF-2 and that the affinity of a given mRNA for eIF-2 can be correlated directly with its ability to compete in translation came from a study of the competition between globin mRNA and mengovirus RNA (Rosen et al. 1982). In conditions where the total number of initiations remains constant, the addition of increasing amounts of mengovirus RNA leads to a progressive decrease in globin mRNA translation, accompanied by increasing synthesis of viral protein (Fig. 3). From a plot of the integrated translation yields (Fig. 4A), it can be calculated that half-maximal inhibition of globin mRNA translation occurs when 35 molecules of globin mRNA are present for every molecule of mengovirus RNA. Assuming that equal proportions of these RNA molecules are translationally active, this means that a molecule of mengovirus RNA competes 35-fold more strongly in translation than does (on average) a molecule of globin mRNA.

Does this competition involve eIF-2? Indeed, in conditions where globin synthesis is greatly depressed by the presence of mengovirus RNA, the addition of eIF-2 does not stimulate overall translation, yet restores globin synthesis to the level seen in the absence of competing mengovirus RNA (Fig. 4B). Globin synthesis in controls lacking mengovirus RNA is not stimulated by the addition of eIF-2 (Fig. 4B, triangles on right). Hence, addition of eIF-2 allows the more weakly competing, but more numerous, globin mRN molecules to initiate translation at the expense of the more strongly competing, but less numerous, viral RNA molecules: eIF-2, therefore, relieves the mRNA competition.

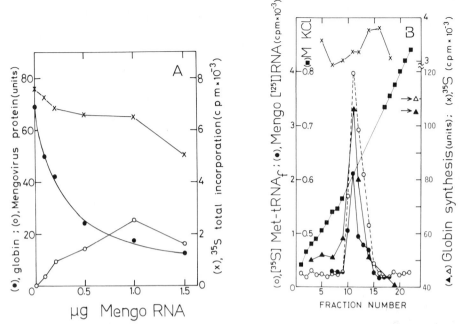

Fig. 4. A Translational competition between globin mRNA and mengovirus RNA. Areas under the curves of densitometer scans (Fig. 3) are plotted in arbitrary units as total amounts of globin (●) and of mengovirus RNA-directed products (○). Total ^{35}S-methionine incorporation into protein (X); **B** Copurification of the activity that relieves translational competition with eIF-2. For purification of eIF-2, see Rosen et al. (1982). The gradient portion of the phosphocellulose column is shown. Aliquots were assayed for binding of ^{35}S-labeled Met-tRNA$_f$ (○) or binding of ^{125}I-labeled mengovirus RNA (●). Aliquots were added to translation mixtures containing 1 μg of globin mRNA and 0.5 μg of mengovirus RNA. Incorporation of ^{35}S-methionine into total protein (X) and total amount of globin formed (▲) are shown. *Triangles on right* indicate amount of globin synthesized in reaction mixtures lacking mengovirus RNA, incubated with (▲) and without (△) material from tube 11. From Rosen et al. (1982)

The fact that eIF-2 acts to shift translation in favor of globin synthesis shows clearly that globin mRNA and mengovirus RNA compete for eIF-2, but does not eliminate the possibility that eIF-2 could act in a nonspecific manner, as by increasing the pool of 40 S/Met-tRNA$_f$ complexes. The results of RNA binding experiments, however, show that mengovirus RNA and globin mRNA compete directly for eIF-2 with an affinity ratio that matches exactly with that observed in translation competition experiments.

In the experiment of Fig. 5, the only macromolecules present are eIF-2 and mRNA. Binding of labeled mengovirus RNA to a limiting amount of eIF-2 is studied in the presence of increasing amounts of unlabeled, competing RNA. Unlabeled mengovirus RNA competes as expected, with the same affinity for eIF-2 as the labeled viral RNA. By contrast, 30 times more molecules of globin mRNA must be present before binding of labeled mengovirus RNA is reduced by one-half. The reciprocal experiment, in which the label is in globin mRNA, yields the same result (Rosen et al. 1982). Thus, a molecule of mengovirus RNA binds to eIF-2 30-fold more strongly than (on average) a molecule of globin mRNA. The high affinity of mengovirus RNA for eIF-2 is not related simply to nucleotide length, for vesicular stomatitis virus (VSV) negative-strand RNA, which

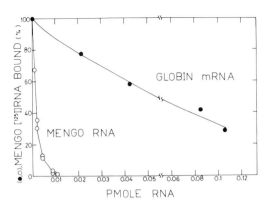

Fig. 5. Competition between globin mRNA and mengovirus RNA in direct binding to eIF-2. See text. From Rosen et al. (1982)

is even longer, binds only very weakly and nonspecifically to eIF-2 and lacks the high affinity binding site found in all mRNA species tested, including the far shorter VSV mRNA (Kaempfer et al. 1978a).

Additional and independent evidence for a preferential interaction of eIF-2 and mengovirus RNA during translation comes from a study of the inhibition of translation by dsRNA (Rosen et al. 1981a). DsRNA blocks initiation by inactivating eIF-2 (Kaempfer 1974; Clemens et al. 1975). Rosen et al. (1981) found that, on the one hand, dsRNA binds with higher affinity than globin mRNA to eIF-2 and inhibits globin mRNA translation, while on the other hand, it binds with lower affinity than mengovirus RNA to eIF-2 and fails to establish translational inhibition when mengovirus RNA is present. The results of that study indicate that the rate-determining event in the establishment of translational inhibition by dsRNA involves competition between mRNA and dsRNA and that mengovirus RNA, because of its higher affinity for eIF-2, is able to protect this initiation factor against inactivation by dsRNA. The implications of these findings for translational control during virus infection have been discussed (Kaempfer et al. 1983).

2.3.2 Competition Between α- and β-Globin mRNA for eIF-2

The addition of increasing amounts of rabbit globin mRNA to the mRNA-dependent reticulocyte lysate generates conditions of increasing mRNA competition pressure that considerably magnifies even small differences in competing ability between individual mRNA species. As seen in Fig. 6A, α- and β-globin are synthesized in about equimolar yield when mRNA is subsaturating ($< 1\ \mu g$), but beyond that point, β-globin synthesis occurs at the progressive expense of α-globin mRNA translation. The addition of a constant amount of purified eIF-2 does not change overall translation at any mRNA concentration (Di Segni et al. 1979), but as seen in Fig. 6B (upper curve), increases the α/β synthetic ratio as compared to the control (lower curve). At low mRNA concentrations, the addition of eIF-2 causes a shift from about 0.8 to about 1.3, approaching the molar ratio of α- to β-globin mRNA (about 1.4-1.5; Lodish 1971). When the relative translation yield equals the molar ratio of the mRNAs under study, there is no competition in translation. Hence, eIF-2 acts to relieve competition. We not here that at very low concentrations of mRNA, the α/β synthetic ratio exceeds 1, and eIF-2 has no ef-

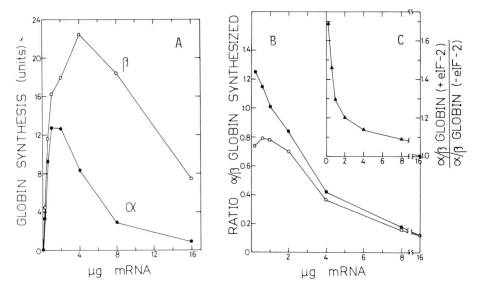

Fig. 6 A-C. A Synthesis of α- and β-globin as a function of globin mRNA concentration; **B** Effect of eIF-2 on the α/β synthetic ratio, computed for the samples incubated without added eIF-2 (○) or for parallel samples that received a constant amount of eIF-2 (●); **C** The effect of eIF-2 on the α/β ratio was determined by dividing the ratio observed in the presence of added eIF-2 by that observed in its absence, and is plotted as a function of the amount of mRNA present. From Di Segni et al. (1979), with permission. Copyright 1979 American Chemical Society

fect, but as the counts are very low in such conditions, the data are not shown (note that the curves do not start at zero mRNA). The relieving effect of eIF-2 is abolished by increasing the mRNA concentration (Fig. 6C).

We have asked if translational competition occurs in a less extensively differentiated tissue than reticulocytes, the liver, that synthesizes a more complex spectrum of proteins, in order to test the prediction that the more evenly different species of mRNA are expressed, the less extreme must be the differences in competing ability between them. Using quantitative immunoprecipitation of products of cell-free translation, the existence of translational competition between individual mRNAs of the liver was demonstrated in similar experiments: the mRNA species encoding hemopexin, ferritin, and albumin possess a progressively greater ability to compete in translation (Kaempfer and Konijn 1983). The differences in competing strength are less pronounced than in the case of α- and β-globin mRNA. Here, too, the more effective template, albumin mRNA, is better able to compete in translation for eIF-2 than is ferritin mRNA, a weaker template.

2.4 Studies with Monovalent Anions

Translation of globin mRNA is inhibited by increasing concentrations of Cl⁻ or OAc⁻, the former ion being more inhibitory (Fig. 7A). These anions primarily inhibit the translation of α-globin mRNA, resulting in a decrease in the α/β globin synthetic ratio (Fig.

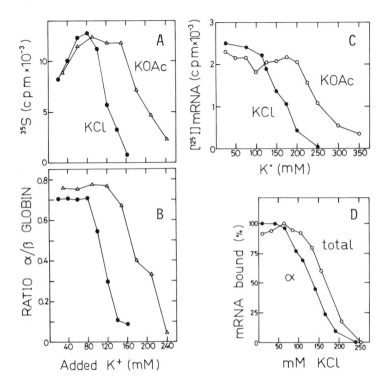

Fig. 7 A-D. A Effect of KCl and KOAc on translation of globin mRNA; **B** on the α/β globin synthetic ratio; **C** on complex formation between ^{125}I-labeled globin mRNA and eIF-2; **D** effect of KCl
on complex formation between eIF-2 and labeled α-globin mRNA or total globin mRNA. From Di
Segni et al. (1979), with permission. Copyright 1979 American Chemical Society

7B). The inhibition of α-globin mRNA translation by Cl⁻ is readily relived by the addition of eIF-2 (Di Segni et al. 1979). Since Cl⁻ or OAc⁻ ions inhibit translation principally by affecting the binding of mRNA to 40 S subunits carrying Met-tRNA$_f$ (Weber
et al. 1977), the relieving activity of eIF-2 suggests that these anions may inhibit the
interaction between eIF-2 and mRNA during translation. Indeed, when the effect of
Cl⁻ or OAc⁻ ions on direct binding of globin mRNA to eIF-2 is studied, a remarkably
parallel inhibition is observed (Fig. 7C).

From the translational competition occuring between α- and β-globin mRNA, and
the relieving effect of eIF-2 thereon (Fig. 6), one would predict β-globin mRNA to
possess a higher affinity for eIF-2 than α-globin mRNA. That this is indeed the case
follows from the results of Fig. 7D. Binding of purified α-globin mRNA to eIF-2 is
more sensitive to KCl than is binding of unfractionated globin mRNA (two-thirds of
which consists of α-globin mRNA; Lodish 1971), attesting to a greater affinity of β-
globin mRNA for eIF-2 (Di Segni et al. 1979).

In Fig. 8, it can be seen that complex formation between mengovirus RNA and eIF-2
continues to take place at KCl concentrations that no longer allow stable binding of
globin mRNA to eIF-2. This result provides independent evidence for the far greater
affinity of mengovirus RNA for eIF-2, already documented in Fig. 5.

Fig. 8. Salt sensitivity of complex formation between eIF-2 and globin mRNA or mengovirus RNA. Binding of ^{125}I-labeled globin mRNA or mengovirus RNA to a limiting amount of eIF-2 was assayed at the indicated concentrations of KCl. From Rosen et al. (1982)

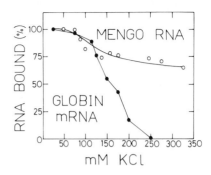

3 Discussion

3.1 eIF-2 as a Target of mRNA Competition

The concept that mRNA species differ in their efficiency of translation, apparently because of a different affinity for one or more critical components in the initiation step, was first suggested by Lodish (1971, 1974) who showed that initiation of protein synthesis on a molecule of α-globin mRNA occurs with lower frequency than that on a molecule of β-globin mRNA. Lawrence and Thach (1974) observed translational competition between host mRNA and encephalomyocarditis virus RNA in a cell-free system from mouse ascites cells and concluded that the competition was at the level of initiation. Hackett et al. (1978a, b) made similar observations for mengovirus RNA. In searching for a possible target of competition, Ray et al. (1983), using a reconstituted cell-free system, suggested that it is initiation factor eIF-4A, or a complex factor, eIF-4F (also named CBP-II), which is able to bind to the 5'-terminal cap structure. The results reported here, showing that eIF-2 is able to relieve translational competition between α- and β-globin mRNA, or between mengovirus RNA and globin mRNA, and that these mRNA species compete directly in binding to eIF-2, cannot be explained by assuming a contamination. First, eIF-2 prepared by our procedure is about 98% pure (Rosen and Kaempfer 1979). Second, the purification procedure yields an eIF-2 preparation that contains only a single mRNA-binding component, eIF-2, since binding of mRNA to this preparation is completely sensitive to competitive inhibition by Met-tRNA$_f$, provided GTP is present (Fig. 1). Third, independent verification that eIF-2 binds 30-fold more tightly to mengovirus RNA is provided by the observation that mengovirus RNA is 30 to 40 times more effective than globin mRNA as a competitive inhibitor of ternary complex formation between Met-tRNA$_f$, GTP, and eIF-2 (Rosen et al. 1981a). While it is conceivable that other initiation factors may also be a target for mRNA competition in translation, it is nevertheless clear from our studies that the addition of eIF-2 is sufficient to overcome such competition.

It is worth noting that in our translation experiments, the added eIF-2 was always used immediately after purification. Storage of purified eIF-2 leads to a loss of activity. Even though stored preparations can be active in ternary complex formation with Met-tRNA$_f$ and GTP, they contain a considerable proportion of inactive initiation factor molecules that can competitively inhibit the active ones. This may explain the failure of other groups to obtain more than marginal relief of translational competition by eIF-2, although the addition of too few molecules of eIF-2 could also explain the results (Ray et al. 1983). Moreover, our translation experiments were done in the microoccal nuclease-treated reticulocyte lysate. This system offers several advantages over recon- stituted cell-free systems. It responds to translational control signals, it is capable of extensive and efficient initiation in conditions more likely to be representative of protein synthesis in intact cells, and except for mRNA, it contains all other components for protein synthesis in a proportion much closer to that of the intact cell.

The ability of eIF-2 to overcome the Cl^- induced inhibition of α-globin mRNA translation (Fig. 7B and Di Segni et al. 1979), coupled with the remarkably parallel effect of Cl^- and OAc^- ions on translation of globin mRNA on the one hand (Fig. 7A) and on binding of globin mRNA to eIF-2 on the other (Fig. 7C), strongly suggest that a direct interaction between mRNA and eIF-2 occurs during protein synthesis. The mRNA competition studies lead to an identical conclusion.

In the translation competition experiments, relief of competition by added eIF-2 was not accompanied by any stimulation of protein synthesis. This explains why, when mengovirus RNA was also present, the increase in globin mRNA translation caused by addition of eIF-2 was concomitant with a decrease in mengovirus RNA translation (Fig. 4B and Rosen et al. 1982). The fact that eIF-2 acts to shift translation in favor of globin synthesis shows clearly that globin mRNA and mengovirus RNA compete for eIF-2, but does not eliminate the possibility that eIF-2 could act in a nonspecific manner, as by increasing the pool of 40 S/Met-tRNAf complexes. The results of the mRNA binding experiments (Figs. 5 and 8), however, showing that mengovirus RNA and globin mRNA compete directly for eIF-2 with an affinity ratio (30-fold) that matches very closely with that observed in translation competition experiments (Figs. 3 and 4A), provide strong evidence for a direct competition of these mRNA species for eIF-2 during trans- lation. While it is conceivable that globin mRNA and mengovirus RNA compete for free eIF-2 molecules, it is more likely that they compete for eIF-2 molecules located in 40 S/Met-tRNA$_f$ complexes (see Di Segni et al. 1979; Rosen et al. 1982, for a fuller discus- sion of this point). The data with α- and β-globin mRNA (Figs. 6 and 7) reinforce these conclusions. Moreover, the greater ability of albumin mRNA to compete in translation, as compared to ferritin and hemopexin mRNA, also appears to be associated with a more favorable competition for eIF-2 (Kaempfer and Konijn 1983).

These results point to mRNA affinity for eIF-2 as a critical element in translational control. They support the concept that translation of different mRNA species is regu- lated to a significant extent by their relative affinities for eIF-2.

3.2 The Site in mRNA Recognized by eIF-2: Role of RNA Conformation

The concept that mRNA competition for eIF-2 leads to differential gene expression becomes more convincing with the finding that eIF-2 by itself recognizes a specific sequence in mRNA and that, at least in the two cases examined, for STNV and mengo- virus RNA, this sequence has turned out to be virtually identical with the ribosome binding site sequence (Kaempfer et al. 1981; Perez-Bercoff and Kaempfer 1982). These results strongly suggest a leading role for eIF-2 in guiding the 40 S ribosomal subunit to its binding site in mRNA. The fact that eIF-2 must be seated on the 40 S subunit before mRNA can be bound fits in well with this concept.

At physiological salt concentrations, complexes between eIF-2 and mRNA are in rapid equilibrium with the free components, while, by contrast, ternary complexes between eIF-2, Met-tRNA$_f$ and GTP are not. This is well-illustrated by the observation that at 100 mM KCl, ternary complex formation can be inhibited over 90% by globin mRNA, while at 280 mM KCl, less than 20% inhibition is observed (Kaempfer et al. 1978b). Hence, free eIF-2 molecules will tend to form ternary complexes rather than bind to mRNA, a stable interaction with mRNA apparently occurring only once GTP and Met-tRNA$_f$ have been bound and eIF-2 has moved to the 40 S subunit.

By nuclease sensitivity analysis, Kaempfer et al. (1981) found that binding of eIF-2 to intact STNV RNA results in a considerable conformational change in mRNA sequences located at, and just downstream from, the ribosome binding site, at positions 44 and about 60 (see Fig. 2A). The observation that the affinity of short, 5'-terminal STNV RNA fragments for eIF-2 increases with nucleotide length (Fig. 2B), even though binding is specific in all cases (Kaempfer et al. 1981), also indicates that features beyond the actually protected sequence may be important for the binding and recognition by eIF-2.

It is quite possible that in addition to the sequence at the ribosome binding site, eIF-2 recognizes RNA conformation features at and around that site. The role of the 5'-leader sequence in determining initiation efficiency has been difficult to assess, in view of a complete lack of unifying features. This point is well-illustrated by the five mRNA species of VSV. *In vivo*, these mRNAs are apparently translated with identical efficiencies of initiation (Villareal et al. 1976), yet they possess 5' leaders that are unrelated in sequence and vary in length from 10 to 41 nucleotides (Rose 1980).

This variation in 5' ends in the face of identical initiation efficiency suggests that it may not be the leader sequence *per se* that determines initiation strength, but the interaction of this sequence with other, internal parts of the RNA molecule. Since the coding sequence in an mRNA molecule is dictated by the individual protein encoded, and since the 3'-untranslated sequence is also highly variable (see Kozak 1983), any stable interaction between 5' leader and internal sequences is only possible if the leader sequence is especially tailored to allow such a fit. If this view is correct, then an important determinant for the efficiency of initiation is the structure generated by this interaction around the initiation codon. The fact that denaturation of eukaryotic mRNA does not lead to binding of ribosomes at internal sites in mRNA (Kozak 1980a; Collins et al. 1982; Zagorska et al. 1982) is not in conflict with this concept. It merely indicates that structure in mRNA is not important for determining *where* initiation occurs. This point should be separated clearly from the question, *how often* initiation takes place. Two mRNA species can vary widely in initiation efficiency, even if, as is generally the case, initiation occurs at the first AUG codon in both. It is here that the contribution of structure may be of essence (Kaempfer 1984).

3.3 The Respective Roles of Various Initiation Factors in the Binding of mRNA to Ribosomes

Sonenberg (1981) observed that cross-linking of the component polypeptides of the cap-binding factor eIF-4F (CBP II) to the cap requires hydrolysis of ATP and suggested that the cap-binding proteins may act to unwind the secondary structure at the 5' end of capped eukaryotic mRNAs. A reduction in 5'-end secondary structure indeed reduces the dependence of mRNA translation on cap-binding proteins (Sonenberg et al. 1982; Lee et al. 1983) and ATP (Morgan and Shatkin 1980; Kozak 1980a, b). Binding of eIF-4A, a component of eIF-4F (Grifo et al. 1983) to mRNA, moreover, is ATP-dependent (Grifo et al. 1982). Curiously, eIF-4A, eIF-4B, and eIF-4F stimulate translation of uncapped STNV RNA to the same extent as that of globin mRNA (Grifo et al. 1983), indicating that, conceivably, they interact with other features in mRNA besides the cap. Indeed, translation of the uncapped polio virus RNA is unusually dependent upon eIF-4A (Daniels-McQueen et al. 1983). In this context, it may be noted that translation of STNV RNA or the uncapped cowpea mosaic virus RNA in the wheat germ system is sensitive to inhibition by cap analogs (Seal et al. 1978). Binding of the uncapped mengovirus RNA to eIF-2, moreover, is also sensitive to inhibition by cap analogs (Kaempfer et al. 1978b). These observations all suggest that recognition of the cap by initiation factors may be part of a more general interaction with the mRNA molecule.

It is tempting to suggest that eIF-4A (Grifo et al. 1982; Ray et al. 1983) and/or cap-binding proteins (Sonenberg 1981; Lee et al. 1983) may unwind secondary structure at the 5' end of mRNA and anchor it at the 5'-terminal cap, while eIF-2 then acts to recognize and bind the sequence and conformation at the ribosome binding site. The conformational change induced in mRNA by binding of eIF-2 (see above), moreover, may create space to accommodate the attached 40 S ribosomal subunit.

Acknowledgment. This work was supported by grants from the National Research Council of Israel, the Gesellschaft für Strahlen- und Umweltforschung (Muenchen) and the Israel Academy of Sciences.

References

Barrieux A, Rosenfeld MG (1977) Characterization of GTP-dependent Met-tRNA$_f$ binding protein, J Biol Chem 252: 3843-3847

Barrieux A, Rosenfeld MG (1978) mRNA-induced dissociation of initiation factor 2. J Biol Chem 253: 6311-6315

Browning KS, Leung DW, Clark JM Jr (1980) Protection of satellite tobacco necrosis virus ribonucleic acid by wheat germ 40 S and 80 S ribosomes. Biochemistry 19: 2276-2282

Chaudhuri A, Stringer EA, Valenzuela D, Maitra U (1981) Characterization of eIF-2 containing two polypeptide chains of Mr = 48,000 and 38,000. J Biochem (Tokyo) 256:3988-3994

Chen Y, Woodley C, Bose K, Gupta NK (1972) Protein synthesis in rabbit reticulocytes: characteristics of a Met-tRNA$_f{}^{Met}$ binding factor. Biochem Biophys Res Commun 48: 1-9

Clemens MJ, Safer B, Merrick WC, Andersen WF, London IM (1975) Inhibition of protein synthesis in rabbit reticulocyte lysates by double stranded RNA and oxidized glutathione: indirect mode of action on polypeptide chain initiation. Proc Natl Acad Sci USA 72: 1286-1290

Collins P, Fuller F, Marcus P, Hightower L, Ball LA (1982) Synthesis and processing of Sindbis virus nonstructural proteins *in vitro*. Virology 118: 363-379

Daniels-McQueen S, Detjen B, Grifo JA, Merrick WC, Thach RE (1983) Unusual requirements for optimum translation of polio viral RNA *in vitro*. J Biol Chem 258: 7195-7199

Darnbrough CH, Legon S, Hunt T, Jackson RJ (1973) Initiation of protein synthesis: evidence for messenger RNA independent binding of methionyl-transfer RNA to the 40 S ribosomal subunit. J Mol Biol 76: 379-403

Di Segni G, Rosen H, Kaempfer R (1979) Competition between α- and β-globin messenger ribonucleic acids for eukaryotic initiation factor 2. Biochemistry 18: 2847-2854

Grifo JA, Tahara SM, Leis JP, Morgan MA, Shatkin AJ, Merrick WS (1982) Characterization of eukaryotic initiation factor 4A, a protein involved in ATP-dependent binding of globin mRNA. J Biol Chem 257: 5246-5252

Grifo JA, Tahara SM, Morgan MA, Shatkin AJ, Merrick WC (1983) New initiation factor activity required for globin mRNA translation. J Biol Chem 258: 5804-5810

Hackett PB, Egberts E, Traub P (1978a) Selective translation of mengovirus RNA over host mRNA in homologous, fractionated, cell-free translational systems from Ehrlich-ascites tumor cells. Eur J Biochem 83: 353-361

Hackett PB, Egberts E, Traub P (1978b) Translation of ascites and mengovirus RNA in fractionated cell-free systems from uninfected and mengovirus-infected Ehrlich-ascites tumor cells. Eur J Biochem 83: 341-352

Hershey JWB (1982) The initiation factors. In: Perez-Bercoff R (ed) Protein biosynthesis in eukaryotes. Plenum, New York, pp 97-117

Jackson RJ (1982) The cytoplasmic control of protein synthesis. In: Perez-Bercoff R (ed) Protein biosynthesis in eukaryotes. Plenum, New York, pp 363-418

Kaempfer R (1974) Identification and RNA binding properties of an initiation factor capable of relieving translational inhibition induced by heme deprivation or double-stranded RNA. Biochem Biophys Res Commun 61: 591-597

Kaempfer R (1979) Purification of initiation factor eIF-2 by RNA-affinity chromatography. In: Grossman L, Moldave K (eds) Methods in enzymology, vol LX, part G. Academic, New York, pp 247-255

Kaempfer R (1984) Regulation of eukaryotic translation. In: Fraenkel-Conrat H, Wagner RR (eds) Comprehensive virology, vol XIX. Plenum, New York, pp 99-175

Kaempfer R, Konijn AM (1983) Translational competition by mRNA species encoding albumin, ferritin, hemopexin and globin. Eur J Biochem 131: 545-550

Kaempfer R, Hollender R, Abrams WR, Israeli R (1978a) Specific binding of messenger RNA and methionyl-tRNA$_f$^{Met} by the same initiation factor for eukaryotic protein synthesis. Proc Natl Acad Sci USA 75: 209-213

Kaempfer R, Rosen H, Israeli R (1978b) Translational control: recognition of the 5' end and an internal sequence in eukaryotic mRNA by the initiation factor that binds methionyl-tRNA$_f$^{Met}. Proc Natl Acad Sci USA 75: 650-654

Kaempfer R, Hollender R, Soreq H, Nudel U (1979a) Recognition of messenger RNA in eukaryotic protein synthesis: equilibrium studies of the interaction between messenger RNA and the initiation factor that binds methionyl-tRNA$_f$. Eur J Biochem 94: 591-600

Kaempfer R, Israeli R, Rosen H, Knoller S, Zilberstein A, Schmidt A, Revel M (1979b) Reversal of the interferon-induced block of protein synthesis by purified preparations of eucaryotic initiation factor 2. Virology 99: 170-173

Kaempfer R, Van Emmelo J, Fiers W (1981) Specific binding of eukaryotic initiation factor 2 to satellite tobacco necrosis virus RNA at a 5'-terminal sequence comprising the ribosome binding site. Proc Natl Acad Sci USA 78: 1542-1546

Kaempfer R, Rosen H, Di Segni G, Knoller S (1983) Structural feature of picornavirus RNA involved in pathogenesis: a very high affinity binding site for a messenger RNA-recognizing protein. In: Kohn A, Fuchs P (eds) Developments in molecular virology, vol VI: mechanisms of viral pathogenesis. Martinus Nijhoff, The Hague, pp 180-200

Kozak M (1980a) Influence of mRNA secondary structure on binding and migration of 40 S ribosomal subunits. Cell 19: 79-90

Kozak M (1980b) Role of ATP in binding and migration of 40 S ribosomal subunits. Cell 22: 459-467

Kozak M (1983) Comparison of initiation of protein synthesis in procaryotes, eucaryotes, and organelles. Microbiol Rev 47: 1-45

Lawrence C, Thach RE (1974) Encephalomyocarditis virus infection of mouse plasmacytoma cells. I. Inhibition of cellular protein synthesis. J Virol 14: 598-610

Lee KAW, Guertin D, Sonenberg N (1983) mRNA secondary structure as a determinant in cap recognition and initiation complex formation. J Biol Chem 258: 707-710

Leung DW, Browning DS, Heckmann JE, RajBhandary UL, Clark JM Jr (1979) Nucleotide sequence of the 5' terminus of satellite tobacco necrosis virus ribonucleic acid Biochemistry 18: 1361-1366

Levin DH, Kyner D, Acs G (1973) Protein initiation in eukaryotes: formation and function of a ternary complex composed of a partially purified ribosomal factor, methionyl transfer RNA, and guanosine triphosphate. Proc Natl Acad Sci USA 70: 41-45

Lin S, Riggs AD (1975) A comparison of *lac* repressor binding to operator and nonoperator DNA. Biochem Biophys Res Commun 62: 704-710

Lodish HF (1971) Alpha and beta globin mRNA: different amounts and rates of initiation of translation. J Biol Chem 246: 7131-7138

Lodish HF (1974) Model for the regulation of mRNA translation applied to haemoglobin synthesis. Nature (Lond) 251: 385-388

Morgan MA, Shatkin AJ (1980) Initiation of reovirus transcription by ITP and properties of m^7I-capped, inosine-substituted mRNAs. Biochemistry 19: 5960-5966

Pelham HRB, Jackson RJ (1976) An efficient mRNA-dependent translation system from reticulocyte lysates. Eur J Biochem 67: 247-256

Perez-Bercoff R, Kaempfer R (1982) Genomic RNA of mengovirus: recognition of common features by ribosomes and eukaryotic initiation factor 2. J Virol 41: 30-41

Ray BK, Brendler TG, Adya S et al. (1983) Role of mRNA competition in regulating translation: further characterization of mRNA discriminatory factors. Proc Natl Acad Sci USA 80: 663-667

Rose JK (1980) Complete intergenic and flanking gene sequences from the genome of vesicular stomatitis virus. Cell 19: 415-421

Rosen H, Kaempfer R (1979) Mutually exclusive binding of messenger RNA and initiator methionyl transfer RNA to eukaryotic initiator factor 2. Biochem Biophys Res Commun 91: 449-455

Rosen H, Knoller S, Kaempfer R (1981a) Messenger RNA specificity in the inhibition of eukaryotic translation by double-stranded RNA. Biochemistry 20: 3011-3020

Rosen H, Di Segni G, Kaempfer R (1982) Translational control by messenger RNA competition for eukaryotic initiation factor 2. J Biol Chem 257: 946-952

Schreier MH, Staehelin T (1973) Initiation of eukaryotic protein synthesis: (Met-tRNA$_f$. 40 S ribosome) initiation complex catalysed by purified initiation factors in the absence of mRNA. Nature New Biol 242: 35-38

Seal SN, Schmidt A, Tomaszewski M, Marcus A (1978) Inhibition of mRNA translation by the cap analogue, 7-methylguanosine-5'-phosphate. Biochem Biophys Res Commun 82: 553-559

Sonenberg N (1981) ATP/Mg^{2+}-dependent cross-linking of cap binding proteins to the 5' end of eukaryotic mRNA. Nucleic Acids Res 9: 1643-1656

Sonenberg N, Guertin D, Lee KAW (1982) Capped mRNAs with reduced secondary structure can function in extracts of poliovirus-infected cells. Mol Cell Biol 2: 1633-1638

Trachsel H, Erni B, Schreier M, Staehelin T (1977) Initiation of mammalian protein synthesis. The assembly of the initiation complex with purified initiation factors. J Mol Biol 116: 755-767

Villareal LP, Breindl M, Holland JJ (1976) Determination of molar ratios of vesicular stomatitis virus induced RNA species in BHK_{21} cells. Biochemistry 15: 1663-1667

von Hippel P, Revzin A, Gross CA, Wang AC (1974) Non-specific DNA binding of genome regulating proteins as biological control mechanism. The *lac* operon: equilibrium aspects. Proc Natl Acad Sci USA 71: 4808-4812

Weber LA, Hickey ED, Maroney PA, Baglioni C (1977) Inhibition of protein synthesis by Cl^-. J Biol Chem 252: 4007-4010

Zagorska L, Chroboczek J, Klita S, Szafranski P (1982) Effect of secondary structure of mRNA on the formation of initiation complexes with prokaryotic and eukaryotic ribosomes. Eur J Biochem 122: 265-269

Regulation of Initiation Factor Activity

H.O. VOORMA[1]

1 Introduction

The sequence of events in the initiation of protein synthesis in eukaryotes has been depicted for a number of years as shown in Fig. 1 (Thomas et al. 1981). The initiator Met-tRNA forms a ternary complex with eukaryotic initiation factor eIF-2 and GTP and binds subsequently to a 40S ribosomal subunit that arises by dissociation of a 80S ribosome in which process two factors, eIF-3 and eIF-4C play a role. The 40S preinitiation complex then attaches to messenger RNA which is mediated by four factors, eIF-4A, eIF-4B, eIF-4E, and eIF-1, ATP hydrolysis being a prerequisite. It has been postulated that the leader sequence of the messenger has to adopt a relaxed structure by melting of the secondary structures before the scanning by the 40S ribosome can take place (Kozak 1978; Grifo et al. 1982; Lee et al. 1983). When the initiation codon AUG is reached, the joining with the 60S occurs, triggering the release of the factors and ensuring that they can be used in a following round of initiation. However, recent studies in several laboratories indicated that another cycle, the eIF-2 cycle, has to be added to this scheme in order to complete the sequence of events in initiation (Amesz et al. 1979; Konieczny and Safer 1983; Siekierka et al. 1981; Panniers and Henshaw 1983; Clemens et al. 1982; Voorma et al. 1983; Safer 1983; Ochoa 1983). Reason for this additional cycle is the fact that eIF-2 is not released as free eIF-2, but as an eIF-2.GDP complex. Since the binding constant of GDP is at least one order of magnitude lower than for GTP, an exchange reaction has been postulated (Konieczny and Safer 1983; Siekierka et al. 1981; Panniers and Henshaw 1983; Clemens et al. 1982).

2 The eIF-2 Cycle

The new factor designated eRF (Voorma et al. 1983), eIF-2B (Safer 1983), and GEF (Ochoa 1983) with Mr = 270 000 comprising five different protein subunits; Mr = 82, 65, 58, 39, and 26 x 10^3, has been isolated in two forms, free and complexed with eIF-2. In Fig. 2 the subunit compositions of both eIF-2.eRF and eIF-2 are given. A direct demonstration of the ultimate effect of this new factor is given in the methionyl-puromycin reaction, in which the reaction is stoichiometric without eRF, whereas in the

[1]   Department of Molecular Cell Biology, University of Utrecht, The Netherlands

Mechanisms of Protein Synthesis
ed. by Bermek
© Springer-Verlag Berlin Heidelberg 1984

EUKARYOTES

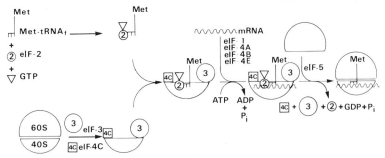

Fig. 1. Initiation of protein synthesis in eukaryotes

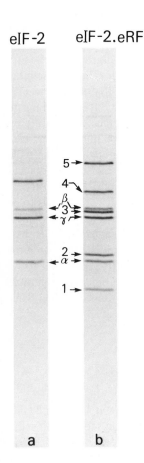

Fig. 2 a, b. Gel analysis of eIF-2 and eIF-2.eRF. 2 μg eIF-2 and 16 μg eIF-2.eRF were analyzed on Laemmli gels (acrylamide: bisacrylamide 30:0.18). α, β, γ: subunits of eIF-2, 1-5: subunits of eRF

Fig. 3. Effect of eRF on the formation of methionyl-puromycin. [³H]Methionyl-puromycin was extracted with ethylacetate saturated with 0.2 M K-phosphate, pH 8.0. ○: no eRF, ●: plus eRF (2 μg)

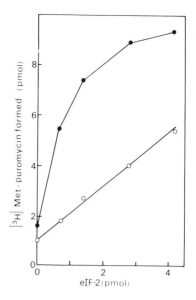

presence of eRF eIF-2 acts catalytically, resulting in the formation of over 3 pmol of methionyl-puromycin per pmol eIF-2 added. The results of this experiment are given in Fig. 3.

To gather insight in the molecular mechanism of this factor a number of experiments were carried out employing the different assay systems that were available. One important difference with normal assays has to be borne in mind; i.e., prior to the incubation, a preincubation step with 3 μM GDP was included, followed by an incubation with increasing GTP concentrations, ranging from 0 to 50 μM GTP. The experiments were carried out in the presence and absence of 40S since earlier observations showed a profound stimulation by eRF when 40S ribosomal subunits were included in the binding experiments (Voorma et al. 1983).

A preliminary experiment in which the GTP-dependence of the ternary complex formation is measured is presented in Fig. 4. One notices a stimulatory effect of eRF on ternary complex formation, which can only be found when no energy-regenerating system of creatine phosphate and creatine kinase is present in the assay mixture. The same experiment, but including 40S subunits is given in Fig. 5. The total amount of bound Met-tRNA is considerably higher than in the previous experiment and the stimulatory effect of eRF has been increased as well. The following set of experiments employed ³H-labeled GDP on the one hand, allowing to study the exchange reaction of GTP with GDP bound to eIF-2, whereas on the other hand, ³H-labeled Met-tRNA has been used in order to follow the binding of Met-tRNA into the ternary complex as well as into the 40S preinitiation complex. The experiments were carried out with both eRF complexed with eIF-2 and with free eRF and eIF-2.

In Fig. 6 the GDP exchange reaction is given. One can conclude from the results of all experiments that GDP readily exchanges with GTP provided the presence of free or complexed eRF. Even at low input of GTP the exchange reaction proceeds to a considerable extent. The counterpart of the exchange experiment is shown in Fig. 7 in

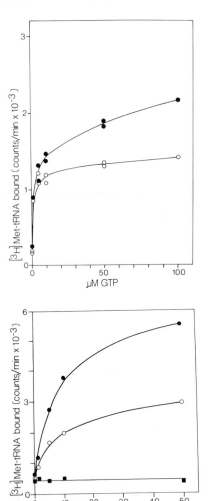

Fig. 4. GTP dependence of ternary complex formation. Ternary complex formation is determined by retention of [^{3}H]Met-tRNA on Millipore filters. ○: eIF-2 (1 μg), ●: eIF-2 (1 μg) and eRF (10 μg)

Fig. 5. GTP dependence of binding of Met-tRNA to 40S. Binding of [^{3}H]Met-tRNA to 40S was determined by Millipore filtration. ○: eIF-2 (1 μg), ●: eIF-2 (1 μg) and eRF (10 μg), ■: eRF (10 μg)

which ^{3}H-labeled Met-tRNA binding is measured under identical experimental conditions. The results of this experiment allow the following conclusions: it is no matter whether complexed or free eRF is used with respect to the stimulatory effect both on ternary complex and on binding to 40S, in all cases the level of binding to 40S, panels B, has been greatly enhanced in the presence of eRF as compared with the level of binding of Met-tRNA in the ternary complex. The result of one such an experiment employing a GTP concentration of 5 μM is given in Fig. 8, establishing the fact that GDP is exchanged very readily and Met-tRNA binding proceeds very well provided that eRF and 40S subunits are taking part in the reaction. The discrepancy to notice is the high GDP exchange and the low ternary complex formation in the presence of eRF. Besides exchange reactions which were carried out until completion we performed a number of kinetic experiments in which the preincubation step with 3 μM GDP has been maintained followed by incubation in the presence of 25 μM GTP.

Fig. 6 a, b. GDP exchange reaction. [³H]GDP, 3 μM, is exchanged with nonlabeled GTP at 0, 1, 5, 10, and 50 μM GTP. Bound GDP is measured by retention on Millipore filters. *Top panel:* **a** ○ eIF-2 (1 μg), ● eIF-2.eRF (2 μg); **b** as **a**, but 10 pmol 40S included. *Bottom panel:* **a** ○ eIF-2 (1 μg), ● eIF-2 (1 μg), eRF (10 μg); **b** as **a**, but 10 pmol 40S included)

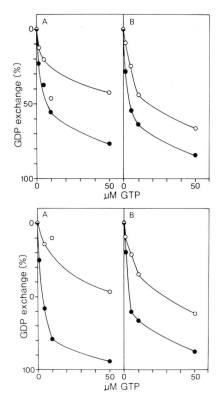

In Fig. 9 the ternary complex formation is measured in panel A, Met-tRNA binding to 40S is given in panel B. The rate of Met-tRNA binding to 40S is much faster, reaching completion in almost 2 min, all the other reactions in which either 40S or eRF were missing display a significant lower rate.

So far the concentration of eRF employed in the exchange and binding reactions was of the same order as the concentration of eIF-2. The question can be raised whether a five times lower concentration of eRF is still capable of achieving a kinetic advantage. In Fig. 10 the results of this experiment are given, showing that the presence of 40S subunits in both reactions, exchange and binding, results in an increase of the rate. The results so far obtained can be explained in two ways, either by assuming that eRF is capable of carrying out an exchange reaction very similar to the prokaryotic exchange reaction with EF-Tu.GDP and EF.Ts (Weissbach et al. 1970), or furthermore that the presence of 40S subunits pulls the reaction to completion because of the formation of stable 40S preinitiation complexes. On the other hand, one may postulate that the complex of eIF-2.eRF binds GTP and Met-tRNA into a quaternary complex, which dissociates into eRF and the ternary complex eIF-2.Met-tRNA.GTP upon binding of the latter to the 40S subunit. Support for the latter possibility can be found from an experiment in which Met-tRNA binding has been studied with the complex eIF-2.eRF and GTP. The reaction mixture has been analyzed by means of glycerol gradient centrifugation and the results indicate the presence of a quaternary complex, sedimenting around 40S, whereas a small amount is present at the tree eIF-2 position, see Fig. 11.

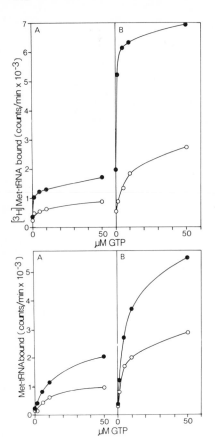

Fig. 7 a, b. Met-tRNA binding into ternary complex and to 40S. The same experiment as in Fig. 6 except for the presence of [³H]Met-tRNA (5 pmol) and omission of [³H]GDP

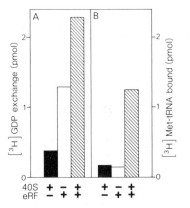

Fig. 8 a, b. Comparison of GDP exchange and Met-tRNA binding. The results of Fig. 6 and Fig. 7 obtained at a GTP concentration of 5 μM have been combined in this figure. 1 pmol Met-tRNA: 4500 counts/min. 1 pmol GDP: 7150 counts/min.

Fig. 9 a, b. Time course of Met-tRNA binding into ternary complex and to 40S subunits. **a** ○ eIF-2 (1 μg), ● eIF − 2.eRF (2 μg); **b** As in **a**, except for the addition of 5 pmol 40S

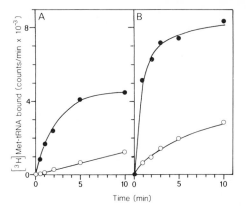

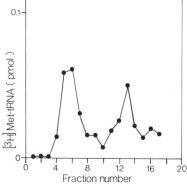

Fig. 10 a, b. Kinetics of Met-tRNA binding to eIF-2 and 40S and GDP-exchange. **a** [³H]GDP 3 μM in preincubation; GTP 25 μM in incubation, ● eIF-2 (1 μg) and eIF-2.eRF (0.5 μg) 40S (5 pmol); ○ without 40S. **b** 3 μM GDP instead of ³H-labeled GDP; [³H]Met-tRNA (5 pmol) instead of nonlabeled Met-tRNA

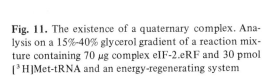

Fig. 11. The existence of a quaternary complex. Analysis on a 15%-40% glycerol gradient of a reaction mixture containing 70 μg complex eIF-2.eRF and 30 pmol [³H]Met-tRNA and an energy-regenerating system

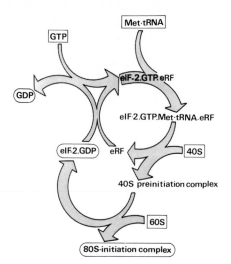

Fig. 12. The eIF-2 and eRF cycle

The presence of the quaternary complex predicts that the eRF moiety is removed during the binding step since no protein subunits of eRF have been found on the 40S preinitiation complex. At the moment eRF is released it becomes available for a subsequent exchange reaction. In Fig. 12 the data presented in this chapter have been combined, indicating the crucial role the 40S subunit plays in the release of eRF. It provides an explanation not only for the higher reaction rate, but also for the higher stimulation that was obtained in the presence of 40S ribosomal subunits.

3 Discussion

The rate of initiation of protein synthesis in eukaryotes is subject to quantitative regulation, the process of which is mainly controlled by the ability of eIF-2 to recycle in subsequent rounds of initiation. The most important prerequisite for undisturbed action of eIF-2 is the eRF-mediated exchange reaction which removes GDP for GTP, thus allowing the formation of a quaternary complex containing eIF-2, eRF, Met-tRNA, and GTP.

The exchange reaction can only be carried out when a complex can be formed between eIF-2 and eRF. Recent studies focusing on the complex formation indicate that covalent modifications of either the α- or β-subunit of eIF-2 may impair the complex formation. The far most studied example is the phosphorylation of the α-subunit of eIF-2 by a protein kinase activated by hemin deficiency which frustrates the complex formation (Voorma and Amesz 1982). Modifications of the β-subunit triggered by metabolic events may become more relevant in the quantitative regulation of initiation. Evidence supporting this view stems from experiments employing oxidized and reduced eIF-2 and ADP-ribosylated eIF-2 (Jagus et al. 1982).

References

Amesz H, Goumans H, Haubrich-Morree T, Voorma HO, Benne R (1979) Purification and characterization of a protein factor that reverses the inhibition of protein synthesis by the heme-regulated translational inhibitor in rabbit reticulocyte lysates. Eur J Biochem 98: 513-520

Clemens MJ, Pain VM, Wong ST, Henshaw EC (1982) Phosphorylation inhibits guanine nucleotide exchange on eukaryotic initiation factor 2. Nature (Lond) 296: 93-95

Grifo JA, Tahara SM, Leis JP, Morgan MA, Shatkin AJ, Merrick WC (1982) Characterization of eukaryotic initiation factor 4A, a protein involved in ATP-dependent binding of globin mRNA. J Biol Chem 257: 5246-5252

Jagus R, Crouch D, Konieczny A, Safer B (1982) The role of phosphorylation in the regulation of eukaryotic initiation factor 2 activity. In: Horecker BL, Stadtman ER (eds) Current topics in cellular regulation, Vol 21, Academic Press, New York, pp 35-63

Konieczny A, Safer B (1983) Purification of the eukaryotic initiation factor 2-eukaryotic initiation factor 2B complex and characterization of its guanine nucleotide exchange activity during protein synthesis initiation. J Biol Chem 258: 3402-3408

Kozak M (1978) How do eucaryotic ribosomes select initiation regions in mRNA. Cell 15: 1109-1123

Lee KAW, Guertin D, Sonenberg N (1983) mRNA secondary structure as a determinant in cap recognition and initiation complex formation. J Biol Chem 258: 707-710

Ochoa S (1983) Regulation of protein synthesis initiation in eukaryotes. Arch Biochem Biophys 223: 325-349

Panniers R, Henshaw EC (1983) A GDP/GTP exchange factor essential for eukaryotic initiation factor 2 cycling in Ehrlich ascites tumor cells and its regulation by eukaryotic initiation factor 2 phosphorylation. J Biol Chem 258: 7928-7934

Safer B (1983) 2B or not 2B: Regulation of the catalytic utilization of eIF-2. Cell 33: 7-8

Siekierka J, Mitsui KI, Ochoa S (1981) Mode of action of the heme-controled translational inhibitor. Relationship of eukaryotic initiation factor 2-stimulating protein to translation restoring factor. Proc Natl Acad Sci USA 78: 220-223

Thomas AAM, Benne R, Voorma HO (1981) Initiation of eukaryotic protein synthesis. FEBS Lett 128: 177-185

Voorma HO, Amesz H (1982) In: Grunberg-Manago M and Safer B (eds) Interaction of translational and transcriptional controls in the regulation of gene expression. Devel. Biochem 24: 297-309. Elsevier Biomedical, New York

Voorma HO, Goumans H, Amesz H, Benne R (1983) In: Horecker BL and Stadtman ER (eds) Current topics in cellular regulation, Vol. 22 Academic Press, New York, pp 51-70

Weissbach H, Miller DL, Hachmann J (1970) Studies on the role of factor T_S in polypeptide synthesis. Arch Biochem Biophys 137: 262-269

Translational Control of Protein Synthesis in Reticulocyte Lysates by eIF-2α Kinases

R.L. MATTS, D.H. LEVIN, R. PETRYSHYN, N.S. THOMAS, and I.M. LONDON[1]

1 Introduction

The initiation of protein synthesis in reticulocyte lysates is inhibited by heme deficiency, double-stranded RNA, or oxidized glutathione (Zucker and Schulman 1968; Rabinovitz et al. 1969; Hunt et al. 1972; Ehrenfeld and Hunt 1971; Kosower et al. 1972). In lysates, the inhibitions are all characterized by a brief period of control linear synthesis, followed by an abrupt decline in this rate as protein synthesis shuts off. The inhibitions are due primarily to the activation of cAMP-independent protein kinases that specifically phosphorylate the 38000 mol. wt. α-subunit of the eukaryotic initiation factor-2 (eIF-2) (Levin et al. 1976; Kramer et al. 1976; Ranu and London 1976; Farrell et al. 1977; Gross and Mendelewski 1977; Levin and London 1978; Ernst et al. 1978a). Similar inhibitions of protein synthesis are produced by adding the purified inhibitors, the heme-regulated eIF-2α kinase (HRI), or the double-stranded RNA-activated eIF-2α kinase (dsI), to reticulocyte lysates. The inhibition of protein synthesis is a result of a block in the early steps of protein chain initiation and is immediately preceded by a decrease in the formation of the [eIF-2·Met-tRNA$_f$·GTP] ternary complex and the [40S·eIF-2·Met-tRNA$_f$·GTP] 43S initiation complex (de Haro et al. 1978; Ranu et al. 1978; de Haro and Ochoa 1978; Ranu and London 1979; de Haro and Ochoa 1979; Das et al. 1979). Studies by Cherbas and London (1976) suggested that the principal effect of HRI was the impairment of the recycling of eIF-2. In vitro studies have suggested that the recycling of eIF-2 requires the presence of an additional multipolypeptide factor, whose function is inhibited by the phosphorylation of eIF-2. This factor whose structure is shown in Fig. 1 has been variously called anti-HRI (Amesz et al. 1979), SP (Siekierka et al. 1982), GEF (Panniers and Henshaw 1983), eIF-2B (Konieczny and Safer 1983), or RF (Siekierka et al. 1981; Matts et al. 1983; Grace et al. 1982). We shall be referring to it here as the reversing factor (RF).

2 Results

2.1 Function of the Reversing Factor (RF)

Restoration of linear synthesis in inhibited lysates can be achieved by the additon of high (nonphysiological) levels of eIF-2 (Kaempfer 1974; Clemens et al. 1975; Ernst et al.

[1] The Harvard Massachusetts Institute of Technology Division of Health Sciences and Technology and The Department of Biology, Cambridge, Massachusetts 02139, USA

Mechanisms of Protein Synthesis
ed. by Bermek
© Springer-Verlag Berlin Heidelberg 1984

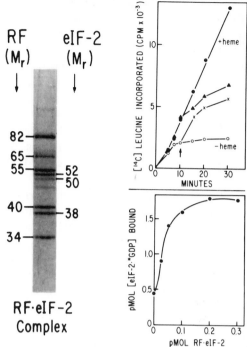

RF
(M$_r$)

eIF-2
(M$_r$)

82—
65—
55— —52
—50
40— —38
34—

RF·eIF-2
Complex

Fig. 1. Two assays for RF: Effect on protein synthesis in heme-deficient lysates and stimulation of binary complex (eIF-2·[^{3}H]GDP) formation. The peak fraction of RF from the glycerol gradient step (Matts et al. 1983) was used in all assays; 2 μg of this preparation analyzed by electrophoresis in SDS-acrylamide (0.1% SDS/10% acrylamide/0.26% BIS) (Levin et al. 1976) and stained with Coomassie blue is shown on the *left*. Except for the β-subunit (50K) of eIF-2, the five subunits of RF and the α (38K) and γ (52K) subunits of eIF-2 appear to be equivalent. *Upper panel:* Protein synthesis assay. Standard lysate protein synthesis assays (30 μl) were carried out for 30 min at 30°C as described (Hunt et al. 1972; Ernst et al. 1978b). •, plus 20 μM hemin-Cl; ○, minus hemin-Cl; ▲, minus hemin-Cl plus 0.5 pmol RF added at 0 min, and at 10 min (*X, arrow*). Added RF stimulated protein synthesis catalytically; approx. 10-20 pmol of globin were synthesized in 30 min for each pmol of RF added. Specific activity of this preparation was 48 000 U/mg of protein where 1 U is the amount of RF required to restore protein synthesis by 50% in heme-deficient lysates. When 1 U is defined as the amount of RF required to stimulate [eIF-2·GDP] complex formation by 50%, then specific activity was 10 000 U/mg protein. *Lower panel:* Binary complex formation. Assays (20 μl) contained 20 mM Tris-HCl (pH 7.6), 100 mM KCl, 250 μM Mg (OAc)$_2$, 0.5 mM dithiothreitol, 100 μg/ml of creatine phosphokinase (as carrier), 3 pmol of eIF-2 (80% pure), 2 μM [^{3}H]GDP (6500 cpm/pmol), and RF as indicated. Incubation was at 30°C for 10 min in Eppendorf tubes (Sarstedt 72-690). The assay was stopped by the addition of 1 ml of cold buffer [20 mM Tris-HCl (pH 7.6)/100 mM KCl/1 mM Mg (OAc)$_2$] and (eIF-2·[^{3}H]GDP) was collected by filtration on Millipore membranes (HAWP 02500) under low vacuum, followed by three 1 ml rinses of each tube and two 3.5 ml washes of the filter. The membranes were heat dried and counted in 5 ml of Econofluor (New England Nuclear). RF acted catalytically; at half-maximal stimulation, 1 pmol of RF catalyzed the formation of 18 pmol of (eIF-2·[^{3}H]GDP). In the absence of added RF, approx. 0.45 pmol of binary complex was formed

1976; or by the addition of approximately ten fold lower levels of RF (Fig. 1, upper right). RF has been purified from both postribosomal supernatants and ribosomal salt washes in either a free form or complexed in a 1:1 stoichiometry with eIF-2 (Amesz et al. 1979; Konieczny and Safer 1983; Siekierka et al. 1981; Matts et al. 1983; Panniers and Henshaw 1983). Addition of RF has been found to stimulate protein synthesis catalytically in inhibited lysates, with 10-20 pmol of globin synthesized in 30 min for each pmol of RF added. Addition of eIF-2 to hemedeficient lysates has been previously found to result in at best a stoichiometric syntheses of globin chains.

Addition of increasing levels of RF to assay mixtures containing eIF-2 and [³H]GDP stimulates the formation of increasing levels of (eIF-2·[³H]GDP) (Fig. 1, lower panel). Earlier studies with similar factors called ESP by de Haro and Ochoa (1978), SF by Ranu and London (1979), and co-eIF-2C by Gupta and colleagues (Das et al. 1979) suggested that these factors stimulated ternary complex formation by interacting with eIF-2 and that this interaction was inhibited by the phosphorylation of eIF-2α. Ranu and London (1979) and Gupta and his colleagues (Das et al. 1979) reported at this

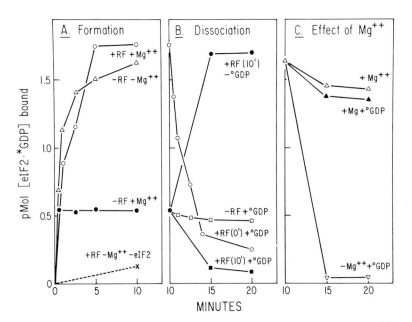

Fig. 2 A-C. Kinetics of [eIF-2·(³H)GDP] formation and dissociation. Role of Mg^{++}. Binary complex assays (20 μl) were carried out with time as described in Fig. 1, except as noted. Each *point* in all curves represents a separate assay. Panel **A** (Formation): •, minus RF; ○, plus 0.2 pmol RF; △, minus RF minus $Mg(OAc)_2$; X, plus 0.2 pmol RF minus eIF-2 minus $Mg(OAc)_2$. Panel **B** (Dissociation): [eIF-2·(³H)GDP] was allowed to form in a standard 10 min incubation at 30°C in the absence of RF (•) (0.55 pmol complex) and in the presence of 0.2 pmol of RF (○) (1.8 pmol complex). At 10 min, additions to both incubations were: □, plus 40 μM unlabeled GDP (°GDP); ■, plus 0.2 pmol RF plus 40 μM °GDP; •, plus 0.2 pmol RF; ○, plus 40 μM °GDP. Panel **C** (Effect of Mg^{++}): [eIF-2·(³H)GDP] was allowed to form (1.7 pmol) at 30°C for 10 min in the absence of both RF and Mg^{++}. At 10 min, additions were: ▽, plus 40 μM °GDP; △, plus 250 μM $Mg(OAc)_2$; ▲, plus 250 μM $Mg(OAc)_2$ plus 40 μM °GDP

time that the presence of this factor was required for the stimulation of ternary complex formation in vitro only in the presence of physiological Mg^{2+} concentrations. Similarly, RF is required for the stimulation of (eIF-2·[^{3}H]GDP) formation only when physiological Mg^{2+} concentrations are present (Fig. 2). In the presence of 250 μM Mg^{2+}, but no added RF, the binding of [^{3}H]GDP to eIF-2 was rapid but limited (0.55 pmol). Further addition of 0.2 pmol of RF increased the binding of [^{3}H]GDP 3.3-fold (Fig. 2A). In the absence of both Mg^{2+} and RF, a similar amount of labeled GDP (1.8 pmol) is bound to eIF-2. This finding, the apparent increase in the binding of [^{3}H]GDP to eIF-2 in the presence of RF, will be discussed later in terms of the mechanism of action of RF.

Addition of excess unlabeled GDP ($^{\circ}$GDP) to the (eIF-2·[^{3}H]GDP) complex formed in the presence of Mg^{2+} alone produced little or no chase of labeled GDP from the complex, but further addition of 0.2 pmol of RF catalyzed a rapid exchange (Fig. 2B). Similarly, the [^{3}H]GDP moiety in the (eIF-2·GDP) complex formed in the presence of both Mg^{2+} and RF undergoes a rapid exchange with added $^{\circ}$GDP. The [^{3}H]GDP moiety in the (eIF-2·GDP) complex formed spontaneously in the absence of both Mg^{2+} and RF (Fig. 2C) is also rapidly chased by excess unlabeled GDP, but this chase can be blocked if Mg^{2+} (250 μM) is added after complex formation, but prior to $^{\circ}$GDP addition. These data support the conclusion that physiological Mg^{2+} concentrations stabilize the (eIF-2·GDP) complex and that RF is required to catalyze the dissociation of bound GDP from eIF-2.

The apparent ability of RF to stimulate the binding of [^{3}H]GDP to eIF-2 in the presence of Mg^{2+} now appears also to be a function of its ability to stimulate (eIF-2 · GDP) dissociation. Siekierka et al. (1983) have recently reported that 50%-100% of the eIF-2 isolated by standard methods from reticulocyte lysates contains bound GDP. In the presence of Mg^{2+}, RF would then be required to dissociate the unlabeled GDP that remained bound to the eIF-2 during its purification. RF-mediated exchange of this unlabeled GDP for [^{3}H]GDP results, therefore, in the apparent stimulation of (eIF-2·[^{3}H] GDP) complex formation in the presence of Mg^{2+}.

2.2 Effect of eIF-2α Phosphorylation on RF Function

The phosphorylation of eIF-2 by HRI results in inhibition of the dissociation of the (eIF-2·GDP) complex (Fig. 3). Phosphorylation of eIF-2 markedly diminishes the extent of RF-mediated (eIF-2·[^{3}H]GDP) formation, as well as RF-catalyzed chase of complex-ed [^{3}H]GDP by unlabeled GDP (Fig. 3A). Phosphorylation of eIF-2 also inhibited the formation of (eIF-2·[^{3}H]GDP) complex formed in the presence of Mg^{2+} alone (Fig. 3C). In addition, phosphorylation of the eIF-2α moiety in the (eIF-2·[^{3}H]GDP) complex blocks the ability of RF to catalyze the exchange with added unlabeled GDP (Fig. 3B). Of interest was the finding that phosphorylated eIF-2 could form a complex with GDP when Mg^{2+} was removed chelation with EDTA (Fig. 3C). Similarly, [^{3}H]GDP in the phosphorylated binary complex is rapidly chased only when Mg^{2+} is removed by EDTA chelation (Fig. 3D). Therefore, under nonphysiological conditions, that is in the absence of Mg^{2+}, GDP can spontaneously dissociate from the (eIF-2 · GDP) complex; as a result, the exchange of [^{3}H]GDP into the complex or the chase of [^{3}H]GDP from labeled

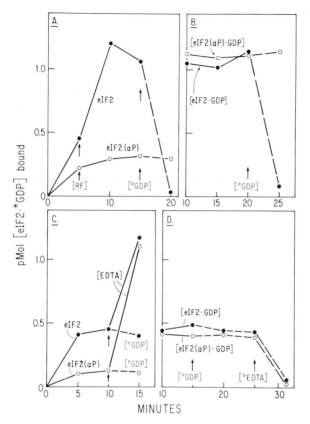

Fig. 3 A-D. Effect of HRI on RF-mediated [eIF-2·(^{3}H)GDP] formation and °GDP exchange. Binary complex formation (in 20 μl) was assayed as described in Fig. 1. Where indicated, eIF-2α phosphorylation, °GDP exchange, and EDTA treatment were carried out in a stepwise manner. Standard incubations (20 μl) contained 20 mM Tris·HCl (pH 7.6), 100 mM KCl, 500 μM Mg(OAc)$_2$, 100 μM creatine phosphate, creatine phosphokinase (100 μg/ml), 0.5 μg of purified HRI (step 7, Matts et al. 1983), and 3 pmol of eIF-2. Other additions and changes were as follows: Panel **A**: Prephosphorylation of eIF-2 in a standard incubation for 10 min plus 40 μM ATP(○) or H$_2$O(●). Subsequent additions to both assays were 2 μM [^{3}H]GDP (6500 cpm/pmol) at 0 min; 0.2 pmol RF at 5 min (*arrows*); 40 μM unlabeled GDP (°GDP) at 15 min (*arrows*). Panel **B**: [eIF-2·(^{3}H)GDP] (1.1-1.2 pmol) was preformed for 10 min in the standard incubation by the addition at 0 min of 2 μM [^{3}H]GDP (6500 cpm/pmol) and 0.2 pmol of RF. Subsequent additions were 40 μM ATP(○) or H$_2$O(●) at 10 min; and 40 μM °GDP at 20 min to both assays (*arrow*). The rate and extent of phosphorylation of eIF-2α in the binary complex was the same as for free eIF-2. Panel **C**: Phosphorylation of eIF-2α in a standard incubation for 10 min plus 40 μM ATP(○) or H$_2$O(●). Subsequent additions to both assays were 2 μM [^{3}H]GDP at 0 min; 40 μM °GDP or 1 mM EDTA at 10 min (*arrows*). Panel **D**: [eIF-2·(^{3}H)GDP] (0.45 pmol) was preformed for 10 min in standard incubations by the addition at 0 min of 2 μM [^{3}H]GDP (6500 cpm/pmol) but no RF. Subsequent additions were 40 μM ATP(○) or H$_2$O(●) at 10 min; 40 μM °GDP to both assays at 15 min (*arrow*); 1 mM EDTA to both assays at 25 min (*arrow*). Controls containing no HRI and no ATP, or containing ATP alone gave results identical to those containing HRI but no ATP, and are not included in the figure

complex is neither dependent on RF nor affected by the phosphorylation of eIF-2α. These data also demonstrate that in the presence of physiological Mg^{2+} concentrations, RF is essential for the dissociation of GDP from the (eIF-2·GDP) complex and that phosphorylation of the binary complex inhibits this effect of RF.

The strong affinity between eIF-2 and GDP probably accounts for the potent inhibition of ternary complex formation by GDP (Walton and Gill 1975; Walton and Gill 1976), and suggests a role for RF in the recycling of eIF-2 in the initiation cycle. Also, it further suggests a mechanism by which eIF-2α phosphorylation inhibits protein synthesis. eIF-2·GDP is presumably formed upon the joining of the 48S complex and the 60S ribosomal subunit. At physiological Mg^{2+} concentrations, in order to recycle eIF-2, GDP in the binary complex must be exchanged for GTP, for which eIF-2 has a much lower affinity than for GDP (Walton and Gill 1975; Walton and Gill 1976). RF catalyzed GTP/GDP exchange would thereby stimulate ternary complex formation in

Table 1. Effect of RF on the conversion of binary complex to ternary complex: Inhibition by HRI

	Additions			
Step 1	Step 2	Step 3	Binary complex	Ternary complex
			(pmol)	(pmol)
eIF-2 + GDP	- - -	- - -	0.31	- - -
eIF-2 + GDP	- - -	Met-tRNAf + GTP	0.31	0.11
eIF-2 + GDP + RF	- - -	- - -	1.30	- - -
eIF-2 + GDP + RF	- - -	Met-tRNAf + GTP	0.13	1.07
eIF-2 + GDP + RF	HRI	- - -	1.28	- - -
eIF-2 + GDP + RF	HRI	Met-tRNAf + GTP	0.36	0.97
eIF-2 + GDP + RF	HRI + ATP	- - -	1.37	- - -
eIF-2 + GDP + RF	HRI + ATP	Met-tRNAf + GTP	1.18	0.18
eIF-2	- - -	Met-tRNAf + GTP	- - -	0.51
eIF-2 + RF	- - -	Met-tRNAf + GTP	- - -	1.20
eIF-2	HRI + ATP	Met-tRNAf + GTP	- - -	0.15
eIF-2	HRI + ATP	Met-tRNAf + GTP + RF	- - -	0.21

Assays were carried out at 30°C in three incubation steps as indicated and complexes were analyzed as described in Fig. 1. *Step 1:* binary complex assays (15 μl) contained 20 mM Tris-HCl (pH 7.6), 100 mM KCl, 500 μM Mg(OAc)$_2$, 0.5 mM dithiothreitol, 100 μg/ml creatine phosphokinase, 100 μM creatine phosphate, 3 pmol eIF-2, and 0.5 μM [^{3}H]GDP (+GDP) in binary complex assays or 0.5 μM unlabeled GDP in ternary complex assays; 0.2 pmol RF (+RF) was added where indicated. Incubation was at 30°C for 10 min. *Step 2:* phosphorylation assay (17 μl) additions were 40 μM ATP and 0.5 μg HRI where indicated. Incubation was at 30°C for 10 min. *Step 3:* ternary complex assay (20 μl) additions were 10 μM GTP and 3 pmol [^{3}H]Met-tRNAf (13,000 cpm/pmol) for ternary complex determinations where indicated. In one set of four ternary complex assays, no labeled or unlabeled GDP was added, but all other components were added at the same concentrations and in a stepwise manner as shown

the presence of Met-tRNAf. The phosphorylation of eIF-2α results in the inhibition of RF-catalyzed exchange of GTP for the GDP bound to eIF-2. This inhibition of exchange results in inhibition of protein synthesis, since GDP will not spontaneously dissociate from eIF-2 in the presence of physiological Mg^{2+} concentrations. Table 1 demonstrates the role of RF in the formation of the ternary complex from the (eIF-2·GDP) complex. Addition of Met-tRNAf and GTP to the eIF-2·GDP complex formed in the presence of Mg^{2+} alone does not change the level of (eIF-2·GDP) complex (0.31 pmol) and produces little ternary complex (0.11 pmol). If RF is also added, greater than 80% of the (eIF-2·GDP) complex formed (1.3 pmol) is converted to ternary complex (1.07 pmol). When the (eIF-2·GDP) complex is phosphorylated by the addition of HRI and ATP, there is a marked inhibition of RF-catalyzed dissociation of the (eIF-2·GDP) complex and its conversion to ternary complex. Therefore, in accord with other studies (Siekierka et al. 1982; Panniers and Henshaw 1983; Matts et al. 1983; Clemens et al. 1982; Pain and Clemens 1983), these findings indicate that even in the presence of GDP, RF promotes ternary complex formation by catalyzing the exchange of GTP for GDP and that this exchange is inhibited upon the phosphorylation of eIF-2α in the (eIF-2·GDP) complex. Since the function of RF in the formation of the ternary complex may only be due to its stimulation of (eIF-2·GDP) dissociation (Siekierka et al. 1983), we have used an (eIF-2·GDP) dissociation assay as an assay for specific RF activity.

2.3 Kinetics of Phosphorylation of eIF-2α Contained in RF

We have examined the effect of RF on the phosphorylation of eIF-2α by HRI. Free eIF-2 was found to be phosphorylated by HRI at a rate ten times that of eIF-2 complexed with RF (Fig. 4). In contrast, eIF-2 in the [RF·eIF-2] complex was found to be rapidly phosphorylated when the complex was added to heme-deficient lysates (London et al. 1983); a finding which suggests that RF does not restore protein synthesis in inhibited lysates merely by protecting eIF-2α from phosphorylation. The rapid phosphorylation of the eIF-2 in the RF·eIF-2 complex in the lysate is explained by the observation that the phosphorylation of the eIF-2α complexed to RF is stimulated six to ten fold in the presence of GDP. This is probably the result of GDP promoted dissociation of [RF·eIF-2], followed by the efficient phosphorylation of the released (eIF-2·GDP) complex.

2.4 Mechanism of Inhibition of RF Activity by eIF-2α Phosphorylation

Phosphorylation of eIF-2 in heme-deficient reticulocytes or their lysates has been found to reach a maximum level around the time at which protein synthesis shuts off (Leroux and London 1982; Farrell et al. 1978). Since only 30%-40% of the eIF-2α present was found to be phosphorylated when translation was inhibited by greater than 90%, it was suggested by Leroux and London (1982) that the functioning of some factor other than eIF-2 (i.e., RF) must become rate limiting in the initiation cycle. On the basis of in vitro studies Siekierka et al. (1982) and Voorma and Amesz (1982) suggested that the phosphorylation of eIF-2 inhibits its interaction with RF thereby causing

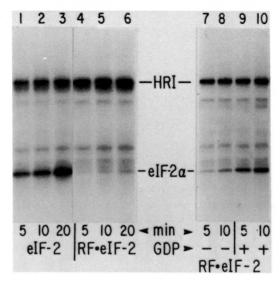

Fig. 4. Autoradiogram of the effect of RF and GDP on eIF-2(α) phosphorylation. In vitro phosphorylation of eIF-2α was carried out using purified components in reaction mixtures (20 μl) containing 10 mM HEPES (pH 7.4), 1 mM Mg(OAc)$_2$, 40 μM [γ-^{32}P]ATP (10-20 Ci/mmol), 50 mM KCl, and 0.05 μg HRI. Other additions were 0.5 pmol eIF-2 (*lanes 1-3*), 0.5 pmol RF (*lanes 4-10*), and 2 μM GDP (*lanes 9-10*). Incubations were carried out at 30°C for the indicated periods and reactions were terminated by the addition of protein dissociation buffer. Samples were heated for 2.5 min in a boiling water bath, subjected to electrophoresis in an SDS polyacrylamide gel (0.1% SDS/10% acrylamide/0.26% BIS), and then stained and dried as described previously (Levin et al. 1976)

inhibition of protein synthesis. However, this hypothesis does not explain why RF would not be free to interact with the 60%-70% of the eIF-2 which remains unphosphorylated in heme-deficient reticulocytes and continue to catalyze protein synthesis. In a reaction mixture containing phosphorylated (eIF-2·GDP) complex, free RF should catalyze the dissociation of subsequently added unphosphorylated (eIF-2·[^{3}H]GDP) complexes. This does not occur, however. The data in Fig. 5 indicate that in the presence of GDP and RF, the phosphorylation of increasing levels of (eIF-2·GDP) leads to the complete inhibition of RF-catalyzed dissociation of unphosphorylated (eIF-2·[^{3}H]GDP). In vitro studies, in which reaction mixtures containing [γ^{32}P]ATP were analyzed by glycerol gradient centrifugation (Fig. 6A), show that in the presence of GDP, RF reacts with phosphorylated (eIF-2(αP)·GDP) complex to form a [RF·eIF-2(αP)] complex. In accord with previous reports (Siekierka et al. 1982; Voorma and Amesz 1982, RF was found not to bind phosphorylated eIF-2 in the absence of GDP (Fig. 6B). Since the concentration of eIF-2 in reticulocyte lysates is estimated to be 10 to 20-fold higher than that of RF, the sequestering of RF in a nonfunctional [RF·eIF-2(αP)] complex, explains how the phosphorylation of as little as 30%-40% of the lysate eIF-2 can inhibit protein synthesis in reticulocyte lysates.

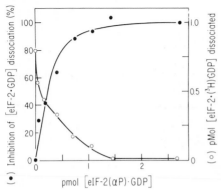

Fig. 5. Inhibition of RF-mediated dissociation of (eIF-2·[³H]GDP) by (eIF-2(αP)·GDP). Reactions were carried out in four steps. *Step 1:* Binary complex assays (8 μl) contained 20 m*M* Tris-HCl (pH 7.6), 100 m*M* KCl, 500 μ*M* Mg(OAc)₂, 2 m*M* dithiothreitol, 100 μg/ml creatine phosphokinase (as carrier), 40 μ*M* ATP, 40 μ*M* GDP, 0.1 pmol RF, and increasing concentrations of eIF-2 (0 to 5.5 pmol). Incubation was 10 min at 30°C. *Step 2:* For phosphorylation of binary complex and free eIF-2 all assays (9 μl) were supplemented with 0.5 μg HRI and incubated for 10 min at 30°C; under these conditions, all free and complexed eIF-2 was phosphorylated. *Step 3:* To remove excess ATP, all assays (11 μl) were supplemented with 0.25 U hexokinase and 5 m*M* glucose and incubated for 7 min at 30°C. Values for the amount of [eIF-2(αP)·GDP] in each assay tube were determined by duplicate assays containing [³H]GDP added in *Step 1* and carried through *Step 3*. *Step 4:* Dissociation of binary complex (21 μl assay) was measured by the addition to each tube of a 10 μl incubation containing excess free eIF-2 (9 pmol) and 1 pmol of unphosphorylated (eIF-2·[³H]GDP) prepared with Mg⁺⁺ but no RF (see Figs. 1, 2). Incubation was 7 min at 30°C. Labeled binary complex was assayed as in Fig. 1. Values are plotted as the percent inhibition of dissociation (●), and as the pmol of (eIF-2·[³H]GDP) dissociated (○) in *Step 4*. Control assays minus HRI (in *Step 2*) carried through all 4 steps yielded maximal dissociation (0.8 pmol) of (eIF-2·[³H]GDP) added in *Step 4*. If RF is omitted in *Step 1*, no dissociation occurs in *Step 4*

2.5 Correlation Between RF Activity and Protein Synthesis in Reticulocyte Lysates

The dissociation of (eIF-2·GDP) is presumably catalyzed specifically by RF in the reticulocyte lysate. We have examined the capacity of whole lysates to catalyze the dissociation of exogenously added (eIF-2·[³H]GDP) to measure their RF activity under various conditions of protein synthesis. In heme-deficient lysates, when protein synthesis is inhibited by over 90%, the rate of (eIF-2·[³H]GDP) dissociation is reduced to 5% of that found in heme-supplemented lysates (Fig. 7). Addition of heme to heme-deficient lysates after the shut-off of protein synthesis restores the rate of protein synthesis and the rate of (eIF-2·[³H]GDP) dissociation to that found in heme-supplemented lysates (Fig. 7). To examine the relationship between RF activity and protein synthesis, we have used the ability of lysates to dissociate (eIF-2·[³H]GDP) in 2 min. as a measure of RF function. Under these conditions the rate of dissociation of (eIF-2·GDP) was found to be linear, such that changes in the amount of (eIF-2·[³H]GDP) dissociated reflect changes in the amount of active RF present.

   Incubation of heme-deficient lysates for various lengths of time results in a progressive loss of the lysates ability to catalyze the dissociation of (eIF-2·[³H]GDP). At 10 min, the loss of the ability of heme-deficient lysates to stimulate (eIF-2·[³H] GDP) dissociation corresponds to the time at which protein synthesis becomes maxim-

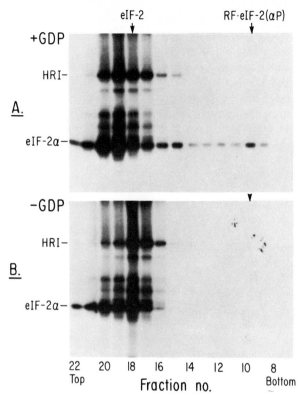

Fig. 6 A, B Formation of a [RF·eIF-2(αP)] complex in the presence of GDP. Reactions were carried out in four steps. *Step 1:* Binary complex incubations (18 µl) contained 20 m*M* Tris-HCl (pH 7.6), 100 µg/ml creatine phosphokinase (as carrier), 500 µ*M* Mg(OAc)$_2$, 2 m*M* dithiothreitol, 20 µ*M* [^{32}P]ATP (10-20 Ci/mmol), 2.5 pmol eIF-2, and 0.6 pmol RF. Incubations were 10 min at 30°C in the presence **A** or absence **B** of 40 µ*M* GDP. *Step 2:* For phosphorylation of eIF-2 all incubations (19 µl) were supplemented with 0.5 µg HRI. Complexed eIF-2 **A** and free eIF-2 **B** were phosphorylated for 10 min at 30°C. *Step 3:* Excess ATP was hydrolyzed by the addition to all assays (21 µl) of 0.25 U of hexokinase and 5 m*M* glucose followed by incubation for 7 min at 30°C. *Step 4:* All assays (22 µl) were supplemented with 5 pmol eIF-2 and incubated for 7 min at 30°C to insure the presence of excess unphosphorylated [eIF-2·GDP] **A** or free eIF-2 **B**. Reaction mixtures were chilled and layered on a 5 ml 15%-50% linear glycerol gradient containing 20 m*M* Tris-HCl (pH 7.5), 100 m*M* KCl, 500 µ*M* Mg(OAc)$_2$, and 2 m*M* dithiothreitol. After centrifugation at 45 000 rpm for 18 h at 2°C in a Beckmann SW50.1 rotor, 0.2 ml fractions were collected and supplemented with carrier creatine phosphokinase (10 µg), ovalbumin (5 µg), and bovine serum albumin (5 µg). Fractions were brought to 0.25 *M* NaOAc (pH 5.5) and protein was precipitated by the addition of 2 vol absolute ethanol. After 20 h at -20°C, precipitates were collected by centrifugation, dissolved in sample buffer, and subjected to electrophoresis, stained and dried as described (Levin et al. 1976). [^{32}P]Phosphoprotein profiles were developed by autoradiography

ally inhibited (Fig. 8). At 7 min, when the rate of protein synthesis is beginning to decline rapidly, heme-deficient lysates show a substantial decrease in (eIF-2·GDP) dissociation activity. Addition of heme to a heme-deficient lysate at 7 min results in a restoration of (eIF-2·GDP) dissociation activity in the lysate which correlates with the restoration of protein synthesis (Fig. 8). Addition of an RF preparation to a heme-deficient lysate at 10 min predictably restores protein synthesis and (eIF-2·[^{3}H]GDP) dissociation activity.

Inhibition ot protein synthesis in lysates by double-stranded RNA or oxidized glutathione is also known to be accompanied by the phosphorylation of eIF-2 (Levin and London 1978; Ernst et al. 1978a). On addition of dsRNA or GSSG, there is a rapid loss of RF activity in the reticulocyte lysate which corresponds to the time of shut-off of protein synthesis (Fig. 9).

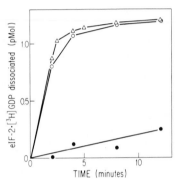

Fig. 7. Kinetics of (eIF-2·[^{3}H]GDP) dissociation in heme-supplemented, heme-deficient, and heme-restored reticulocyte lysates. RF activities in rabbit reticulocyte lysates were measured by their ability to stimulate the dissociation of added (eIF-2·[^{3}H]GDP) binary complex. Reticulocyte lysates were incubated under conditions for protein synthesis essentially as described by Pelham and Jackson (1976). Standard protein synthesis mixtures (100 μl) contained 65.4 μl of rabbit reticulocyte lysate (1:1 lysate prepared as described by Hunt et al. 1972), 80 mM KCl, 0.5 mM MgCl$_2$, 4 μg creatine kinase, 8 mM creatine phosphate, 20 amino acids at 40 μM each, 0.5 mM dithiothreitol, 0.5 mM Mg^{++}GTP, and 20 μM hemin-Cl as indicated. Lysate dilution buffer contained 40 mM Tris HCl (pH 7.6), 100 mM KCl, 50 mM KF, 2 mM Mg(OAc)$_2$, 10% glycerol, and 40 μM unlabeled GDP. KF was present to prevent any possible release of RF due to dephosphorylation of eIF-2α. GDP was added to insure the presence of excess unlabeled nucleotide to exchange with [^{3}H]GDP from dissociating binary complex. (eIF-2·[^{3}H]GDP) was preformed (10 min, 30°C) in a reaction mixture containing 20 mM Tris-HCl (pH 7.6), 100 mM KCl, 2 mM dithiothreitol, 100 μg/ml creatine kinase as carrier, 2 μM [^{3}H]GDP (6500 cpm/pmol), and eIF-2 (~80% pure). Preformed (eIF-2·[^{3}H]GDP) complexes were stored on ice prior to use.

Protein synthesis reaction mixtures (700 μl, 30°C) were diluted with an equal volume of lysate dilution buffer (30°C), followed by the immediate addition of 20 μl, containing 9 pmol of preformed (eIF-2·[^{3}H]GDP) complex. Undissociated (eIF-2·[^{3}H]GDP) remaining in each 200 μl aliquot of diluted protein synthesis mixtures was measured at the times indicated by the retention of the complex on Millipore filters (HAWP 02500). The reaction in each 200 μl aliquot was stopped by the addition of 3 ml of ice cold wash buffer containing 20 mM Tris-HCl (pH 7.6), 100 mM KCl, and 1 mM Mg(OAc)$_2$ and (eIF-2·[^{3}H]GDP) was collected by filtration on Millipore filters. Each tube was then rinsed three times with 3 ml of wash buffer, followed by two 3 ml washed of each filter. Filters were heat dried and radioactivity was determined in 5 ml of Econoflour (New England Nuclear). The total pmol of (eIF-2·[^{3}H]GDP) added was measured by immediately pouring a diluted assay mixture, containing heme-deficient lysate preincubated for 15 min onto a Millipore filter upon addition of the (eIF-2·[^{3}H]GDP) complex. △, Standard protein synthesis mixture incubated for 15 min with 20 μM hemin Cl added at 0 min; •, standard protein synthesis mixture incubated for 15 min without added hemin Cl; ○, standard protein synthesis mixture incubated 30 min with 20 μM hemin Cl added at 10 min

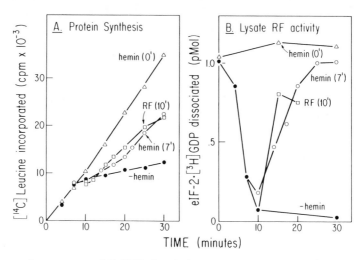

Fig. 8 A, B. Restoration of lysate (eIF-2·[³H]GDP) dissociation activity upon reversal of protein synthesis inhibition by the addition of hemin or RF. **A** Standard lysate-protein synthesis assays (50 μl) containing [¹⁴C]leucine (148 mCi/mmol) were carried out for 30 min at 30°C as described (Hunt et al. 1972; Ernst et al. 1978b). △, with 20 μM hemin Cl added at 0 min; •, without hemin Cl; ○, with 20 μM hemin Cl added at 7 min; ▢, without hemin Cl but with 1 pmol of RF added at 10 min. **B** RF activity was determined in standard protein synthesis mixture similarly incubated at 30°C in the absence of [¹⁴C]leucine. At the times indicated 100 μl aliquots were diluted with an equal volume of lysate dilution buffer, 20 μl containing 1.2 pmol of preformed (eIF-2·[³H]GDP) was immediately added, and the [³H]GDP remaining bound to eIF-2 after 2 min was determined by retention on Millipore filters as described in Fig. 7. △, with 20 μM hemin Cl added at 0 min; •, without hemin Cl; ○, with 20 μM hemin added at 7 min; ▢, without hemin Cl but with 2 pmol of RF added per 100 μl of protein synthesis mix at 10 min.

The RF used in these experiments was a crude preparation purified from a reticulocyte ribosomal salt wash. The 0.2–0.4 M KCl eluant from the phosphocellulose step of the eIF-2 preparation was dialyzed against standard buffer (20 mM Tris HCl, pH 7.5/1 mM dithiothreitol/0.2 mM EDTA/10% glycerol) containing 0.1 M KCl. This preparation was applied to a Sepharose 6B-heparin column (1 ml) previously equilibrated with the above buffer. The column was washed with standard buffer containing 0.3 M KCl and RF activity was eluted with buffer containing 0.5 M KCl. RF was precipitated by addition of $(NH_4)_2SO_4$ to a concentration of 80%. The RF was resuspended in standard buffer containing 0.1 M KCl and the remaining $(NH_4)_2SO_4$ was removed by dialysis

The loss of RF activity in the reticulocyte lysate, measured as (eIF-2·[³H]GDP) dissociation activity, correlates with the known activation of eIF-2α kinases (Levin et al. 1976; Kramer et al. 1976; Ranu and London 1976; Farrell et al. 1977; Gross and Mendelewski 1977; Levin and London 1978; Ernst et al. 1978a), the phosphorylation of eIF-2α, and the inhibition of protein synthesis (Zucker and Schulman 1968; Rabinovitz et al. 1969; Hunt et al. 1972; Ehrenfeld and Hunt 1971; Kosower et al. 1972). At the time when protein synthesis becomes inhibited, the lack of significant (eIF-2·GDP) dissociation activity demonstrates that this function is specific for RF and that RF has become nonfunctional or sequestered.

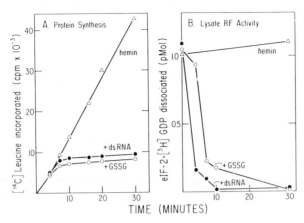

Fig. 9 A, B. Effect of double-stranded RNA and oxidized glutathione on protein synthesis and lysate RF activity. **A** Protein synthesis assay was carried out as described in Fig. 8A. △, with 20 μM hemin Cl added at 0 min; ●, with 20 μM hemin Cl and double-stranded RNA (20 ng/ml) added at 0 min; ○, with 20 μM hemin and 500 μM oxidized glutathione added at 0 min; **B** RF activity was determined in lysates incubated as above and described in Fig. 8B

3 Conclusions

We present the following scheme for RF function and the mechanism by which eIF-2α kinases bring about the inhibition of protein synthesis. (1) The (eIF-2·GDP) binary complex is formed during the joining of the 48S complex and the 60S ribosomal subunit. (2) The critical role of RF is to catalyze the dissociation of (eIF-2·GDP) to permit formation of the (eIF-2-Met-tRNA$_f$·GTP) ternary complex. (3) (eIF-2·GDP) is the primary site of action of eIF-2α kinases. (4) RF interacts with phosphorylated (eIF-2·GDP) complex to form a [RF·eIF-2(αP)] complex that effectively sequesters RF so that it cannot function in the recycling of unphosphorylated eIF-2 which accumulates as an (eIF-2·GDP) complex upon the completion of an initiation cycle. Because RF is present in lysates in a limiting concentration relative to eIF-2, the phosphorylation of 30%-40% of the eIF-2 in heme-deficient cells and lysates suffices to render RF unavailable or nonfunctional and inhibit protein synthesis.

Supported in part by USPHS Grants AM-16272 and GM-24825

References

Amesz H, Goumans H, Haubrich-Morre T, Voorma HO, Benne R (1979) Purification and characterization of a protein factor that reverses the inhibition of protein synthesis by the heme-regulated translational inhibitor in rabbit reticulocyte lysates. Eur J Biochem 98: 513-520

Cherbas L, London IM (1976) Mechanism of delayed inhibition of protein synthesis in heme-deficient rabbit reticulocyte lysates. Proc Natl Acad Sci USA 73: 3506-3510

Clemens MJ, Pain VM, Wong S, Henshaw EC (1982) Phosphorylation inhibits guanine nucleotide exchange on eukaryotic initiation factor 2. Nature (Lond) 296: 93-95

Clemens MJ, Safer B, Merrick WC, Anderson WF, London IM (1975) Inhibition of protein synthesis in rabbit reticulocyte lysates by double stranded RNA and oxidized glutathione: independent mode of action on polypeptide chain initiation. Proc Natl Acad Sci USA 72: 1286-1290

Das A, Ralston RO, Grace M et al. (1979) Protein synthesis in rabbit reticulocytes-mechanism of protein synthesis inhibition by heme-regulated inhibitor 24. Proc Natl Acad Sci USA 76: 5076-5080 deHaro C, Datta A, Ochoa S (1978) Mode of action of the hemin controlled inhibitor of protein synthesis. Proc Natl Acad Sci USA 75: 243-247

deHaro C, Ochoa S (1978) Mode of action of hemin-controlled inhibitor of protein synthesis-studies with factors from rabbit reticulocytes 2. Proc Natl Acad Sci USA 75: 2713-2716

deHaro C, Ochoa S (1979) Further studies on the mode of action of heme controlled translational inhibitor. Proc Natl Acad Sci USA 76: 1741-1745

Ehrenfeld E, Hunt T (1971) Double-stranded poliovirus RNA inhibits initiation of protein synthesis by reticulocyte lysates. Proc Natl Acad Sci USA 68: 1075-1078

Ernst V, Levin DH, London IM (1978a) Inhibition of protein synthesis initiation by oxidized glutathione-activation of a protein kinase that phosphorylates alpha-subunit of eukaryotic initiation factor 2. Proc Natl Acad Sci USA 75: 4110-4114

Ernst V, Levin DH, London IM (1978b) Evidence that glucose 6-phosphate regulates protein synthesis initiation in reticulocyte lysates. J Biol Chem 253: 7163-7162

Ernst V, Levin DH, Ranu RS, London IM (1976) Control of protein synthesis in reticulocyte lysates-effects of 3'-5'-cyclic AMP, ATP and GTP upon inhibitions induced by heme deficiency, double-stranded RNA, and a reticulocyte translational inhibitor. Proc Natl Acad Sci USA 73: 1112-1116

Farrell P, Balkow J, Hunt T, Jackson RJ, Trachsel H (1977) Phosphorylation of initiation factor eIF-2 and the control of reticulocyte protein synthesis. Cell 11: 187-200

Farrell PJ, Hunt T, Jackson RJ (1978) Analysis of phosphorylation of protein synthesis initiation factor eIF-2 by two-dimensional gel electrophoresis. Eur J Biochem 89: 517-521

Grace M, Ralston RO, Banerjee AC, Gupta NK (1982) Protein synthesis in rabbit reticulocytes. Characteristics of the protein factor RF that reverses inhibition in heme-deficient reticulocyte lysates. 30. Proc Natl Acad Sci USA 79: 6517-6521

Gross M, Mendelewski J (1977) Additional evidence that the hemin controlled translational repressor from rabbit reticulocytes is a protein kinase. Biochem Biophys Res Commun 74: 559-569

Hunt T, Vanderhoff G, London IM (1972) Control of globin synthesis, the role of heme. J Mol Biol 66: 471-481

Kaempfer R (1974) Identification and mRNA-binding properties of an initiation factor capable of relieving translational inhibition induced by heme deprivation or double-stranded RNA. Biochem Biophys Res Commun 61: 591-597

Konieczny A, Safer B (1983) Purification of the eukaryotic initiation factor 2, eukaryotic initiation factor 2B complex and characterization of its guanine nucleotide exchange activity during protein synthesis initiation. J Biol Chem 258: 3402-3408

Kosower NS, Vanderhoff GA, Kosower EM (1972) Glutathione VIII. The effects of glutathione disulfide on initiation of protein synthesis. Biochim Biophys Acta 272: 623-627

Kramer G, Cimadivella M, Hardesty B (1976) Specificity of the protein kinase activity associated with the hemin controlled repressor of rabbit reticulocyte. Proc Natl Acad Sci USA 73: 3078-3082

Leroux A, London IM (1982) Regulation of protein synthesis by phosphorylation of eukaryotic initiation factor 2-alpha in intact reticulocytes and reticulocyte lysates. Proc Natl Acad Sci USA 79: 2147-2151

Levin DH, London JM (1978) Regulation of protein synthesis-activation by double-stranded RNA of a protein kinase that phosphorylates eukaryotic initiation factor 2. Proc Natl Acad Sci USA 75: 1121-1125

Levin DH, Ranu RS, Ernst V, London IM (1976) Regulation of protein synthesis in reticulocyte lysates: Phosphorylation of methionyl-tRNA binding factor by protein kinase activity of translational inhibitor isolated from heme-deficient lysate. Proc Natl Acad Sci USA 73: 3112-3116

London IM, Fagard R, Leroux A, Levin DH, Matts R, Petryshyn R (1983) Goldwasser E (ed) Regulation of hemoglobin biosynthesis. Elsevier Biomedical, New York, pp 165-183

Matts RL, Levin DH, London IM (1983) Effect of phosphorylation of the alpha-subunit of eukaryotic initiation factor 2 on the function of reversing factor in the initiation of protein synthesis. Proc Natl Acad Sci USA 80: 2559-2563

Pain VM, Clemens MJ (1983) Assembly and breakdown of mammalian protein synthesis initiation complexes: regulation by guanine nucleotides and by phosphorylation of initiation factor eIF-2. Biochemistry 22: 726-733

Panniers R, Henshaw EC (1983) A GDP/GTP exchange factor essential for eukaryotic initiation factor 2 cycling in Ehrlich ascites tumor cells and its regulation by eukaryotic initiation factor 2 phosphorylation. J Biol Chem 258: 7928-7934

Pelham HRB, Jackson RJ (1976) An efficient mRNA-dependent translation system from reticulocyte lysates. Eur J Biochem 67: 247-256

Rabinovitz M, Freedman ML, Fisher JM, Maxwell CR (1969) Translational control in hemoglobin synthesis. Cold Spring Harbor Symp Quant Biol 34: 567-578

Ranu RS, London IM (1976) Regulation of protein synthesis in rabbit reticulocyte lysates-purification and initial characterization of cyclic 3'-5'-AMP independent protein kinase of heme-regulated translational inhibitor (Phosphorylation of Met-tRNA$_f$ binding factor). Proc Natl Acad Sci USA 73: 4349-4353

Ranu RS, London IM (1979) Regulation of protein synthesis in rabbit reticulocyte lysates-additional initiation factor required for formation of ternary complex (eIF-2·GTP-met-tRNA$_f$) and demonstration of inhibitory effect of heme-regulated protein kinase. Proc Natl Acad Sci USA 76: 1079-1083

Ranu RS, London IM, Das A et al. (1978) Regulation of protein synthesis in rabbit reticulocyte lysates by heme-regulated protein kinase – inhibition of interaction of met-tRNA$_f^{met}$ binding factor with another initation factor in formation of met-tRNA$_f^{met}$. 40S ribosomal subunit complexes. Proc Natl Acad Sci USA 75: 745-749

Siekierka J, Manne V, Mauser L, Ochoa S (1983) Polypeptide chain initiation in eukaryotes. Reversibility of the ternary complex-forming reaction. Proc Natl Acad Sci USA 80: 1232-1235

Siekierka J, Mauser L, Ochoa S (1982) Mechanism of polypeptide chain initiation in eukaryotes and its control by phosphorylation of the alpha-subunit of initiation factor 2. Proc Natl Acad Sci USA 79: 2537-2540

Siekierka J, Mitsui K, Ochoa S (1981) Mode of action of the heme-controlled translational inhibitor: relationship of eukaryotic initiation factor 2-stimulating protein to translation restoring factor. Proc Natl Acad Sci USA 78: 220-223

Walton GM, Gill GN (1975) Nucleotide regulation of eukaryotic protein synthesis initiation complex. Biochim Biophys Acta 390: 231-245

Walton GM, Gill GN (1976) Regulation of ternary (Met-tRNA$_f$.GTP.eukaryotic initiation factor 2) protein synthesis initiation complex formation by the adenylate energy charge. Biochim Biophys Acta 418: 195-203

Voorma HO, Amesz H (1982) Grunberg-Manago M, Safer B (eds) Developments in biochemistry, vol 24. Elsevier Biomedical, New York, pp 297-309

Zucker WV, Schulman HM (1968) Stimulation of globin-chain initiation by hemin in the reticulocyte cell-free system. Proc Natl Acad Sci USA 59: 582-589

Calcium Ions and Phospholipid Activate a Translational Inhibitor in Reticulocyte Lysates

C. DE HARO, A. G. DE HERREROS, and S. OCHOA[1]

1 Introduction

Protein synthesis in rabbit reticulocyte lysates starts at a high rate but declines sharply within a few minutes unless the system is supplemented with hemin (London et al. 1976; Hunt et al. 1972; Mathews et al. 1973). Heme deficiency activates an inhibitor of protein synthesis initiation (heme controlled translational inhibitor or HCI) (Maxwell et al. 1971; Gross and Rabinovitz 1972), a cyclic AMP-independent protein kinase that specifically phosphorylates the small, α subunit of the initiation factor eIF-2 interfering with its function (Farrell et al. 1977; Levin et al. 1976; Kramer et al. 1976; Gross and Mendelewski 1977). There are a number of ways in which HCI can be activated (Ochoa 1983), including high hydrostatic pressure, elevated temperatures, sulfhydryl reagents (e.g., N-ethylmaleimide), or low levels of oxidized glutathione (GSSG), although the mechanism of activation is unknown in all cases. Only the activation due to heme deficiency or GSSG would appear to be physiologically relevant. Activation of HCI in lysates leads to phosphorylation of a polypeptide of $M_r \sim 90\,000$ as well as of the eIF-2 α subunit. The 90 000 M_r polypeptide is HCI itself.

We have observed that calcium ions inhibit protein synthesis in hemin-supplemented reticulocyte lysates and promote phosphorylation of both the eIF-2 α subunit and the polypeptide which comigrates with the 90 000 mol. wt. band of HCI. At 30°C translational inhibition by Ca^{2+} is not reversed by EGTA after 15 min of incubation, but at 20°C the inhibition is fully reversed, even after incubation for 90 min, and the phosphorylation of the eIF-2 α subunit is correspondingly decreased (de Herreros et al. 1983). The fact that the Ca^{2+} promoted activation of HCI is largely prevented by phenothiazines, suggested the involvement of calmodulin in this inhibition. However, calmodulin antibody had no effect on the inhibition produced by Ca^{2+} and authentic calmodulin did not affect translation even in the presence of limiting amounts of Ca^{2+} (de Herreros et al. 1983). The translational inhibition induced by Ca^{2+} is prevented or decreased not only by EGTA, but also by polymyxin B, a bacterial antibiotic that inhibits the activity of Ca^{2+} and phospholipid-dependent protein kinases (Mazzei et al. 1982). Moreover, a 10:1 mixture of phosphatidylserine and diacyl-glycerol (1,3 diolein) inhibits translation and promotes phosphorylation of the eIF-2 α subunit in hemin-supplemented reticulocyte lysates. The results are consistent with the notion that Ca^{2+}

[1] Centro de Biologia Molecular, CSIC-UAM, Madrid-34, Spain, and Roche Institute of Molecular Biology, Roche Research Center, Nutley, New Jersey 07110, USA

Mechanisms of Protein Synthesis
ed. by Bermek
© Springer-Verlag Berlin Heidelberg 1984

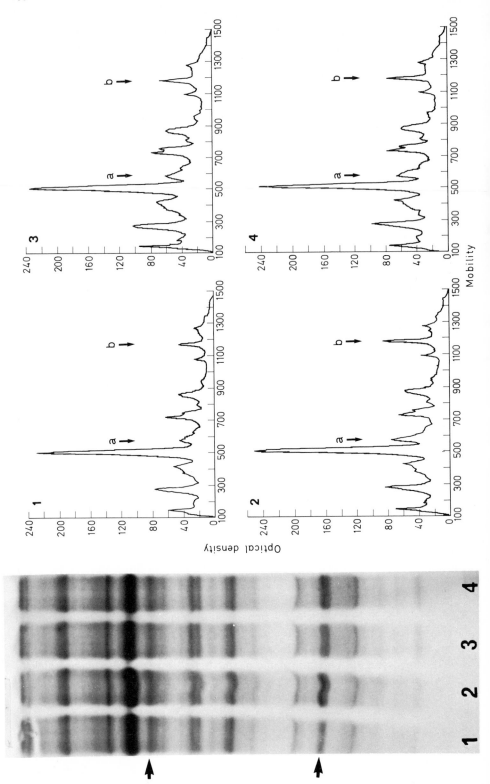

and phospholipid promote directly or indirectly the activation of the proinhibitor form of HCI (de Haro et al. 1983).

Conditions for studying the activation of HCI by Ca^{2+} and phospholipid were according to published procedures (Hunt et al. 1972; de Haro et al. 1983; Jagus and Safer 1981).

2 Translational Inhibition by Ca^{2+}

As seen in Table 1, Ca^{2+} is a potent inhibitor of translation in hemin-supplemented reticulocyte lysates. The extent of inhibition varies widely with different lysates. In general, lysates that are poorly stimulated by hemin addition, are more sensitive to the inhibitory effect of calcium. Under these conditions addition of EGTA markedly stimulates translation in the presence of hemin, an effect that may be due to the presence of relatively high levels of endogenous Ca^{2+} in these lysates. Like the translational in-

Table 1. Effect of Ca^{2+} and EGTA on translation

Experiment No.	Additions[a]	Hemin-dependent translation, cpm x 10^{-3}	Activity %
1	None	30.6	100
	Ca^{2+}, 0.05 mM	20.4	67
	Ca^{2+}, 0.1 mM	8.8	29
	Ca^{2+}, 0.2 mM	2.5	8
	EGTA, 0.1 mM	42.4	138
2	None	9.0	100
	Ca^{2+}, 0.05 mM	4.2	47
	Ca^{2+}, 0.1 mM	1.2	13
	Ca^{2+}, 0.2 mM	0.4	4
	EGTA, 0.1 mM	24.0	267

[a]Incorporations in the control tubes were: (a) with hemin, 38 300 cpm; (b) without hemin, 7 700 cpm; (c) with hemin and EGTA, 51 500 cpm, and (d) with EGTA, 9 100 cpm. Expt. No. 1 and 14 600, 5 600, 31 300, and 7 300 cpm, respectively, in Expt. No. 2. Ca^{2+} was adaed as $CaCl_2$

Fig. 1. Effect of Ca^{2+} on phosphorylation of the 90 000 mol. wt. band of HCl and the eIF-2 α subunit in reticulocyte lysates. The assay was as reported (de Haro et al. 1983). *Left* autoradiogram, *Tracks: 1*, +hemin; *2*, -hemin; *3*, +hemin +Ca^{2+} (0.3 mM); *4*, +hemin +Ca^{2+} (0.6 mM). *Arrows* show the position of the 90 000 mol. wt (*upper*) and 38 000 mol. wt. (*lower*) bands of HCl and the eIF-2 subunit, respectively. *Right* scanning of autoradiogram. *Panel numbers* 1-4 correspond to track numbers in the autoradiogram. *Arrows a* and *b* show the position of the 90 000 and 38 000 mol. wt. bands, respectively. The corresponding 60 min translation values were 78 500, 15 700, 33 000, and 21 200 cpm

hibition due to heme deficiency (see London et al. 1976; Ochoa 1983; Ochoa and de Haro 1979 for reviews), Ca^{2+} inhibition is largely prevented by high levels of added eIF-2 or GTP (data not shown).

We find that the Ca^{2+} promoted translational inhibition is accompanied by increased phosphorylation of both the 38 000 mol. wt. subunit of eIF-2 and a polypeptide that comigrates with the 90 000 mol. wt. band of HCI. The phosphorylation of the 38 000 mol. wt. subunit of eIF-2 is comparable to that seen in the absence of added hemin (Fig. 1). Increased phosphorylation of the 90 000 mol. wt. polypeptide is more clearly seen in the autoradiogram scanning shown in this figure. These results suggest that in the presence of hemin, Ca^{2+} promotes in some way the activation of HCI.

The inhibition by Ca^{2+} is not reversed by EGTA after 15 min of incubation at 30°C and is only partially reversed after 5 min (Fig. 2A). However, at 20°C the inhibition is fully reversed even after incubation for 90 min (Fig. 2B). Consonant with these results, the Ca^{2+}-promoted activation of HCI in reticulocyte lysates is reversed by EGTA at 20°C, but not at 30°C (Fig. 3). It may also be noted that whereas Ca^{2+} promotes HCI activation (translational inhibition) at 20° and 30°C, this is not the case with heme deficiency which is devoid of a translational effect at the lower temperature (Fig. 3A and B). Our results are consistent with the view that Ca^{2+} may activate pro-HCI, directly or indirectly, by promoting the first step of the reaction pro HCI $\rightleftharpoons$ reversible HCI $\longrightarrow$ irreversible HCI.

The Ca^{2+}-promoted phosphorylation of the 38 000 mol. wt. subunit of eIF-2 is partially prevented by phenothiazines. However, calmodulin antibody had no effect on the inhibition produced by Ca^{2+} (Table 2) and highly active calmodulin did not affect translation even in the presence of limiting amounts of Ca^{2+} (Table 3).

3 Translational Inhibition by Phospholipid

The Ca^{2+}-promoted phosphorylation of the eIF-2 α subunit might be due to one of the following causes: (a) limited proteolysis by a Ca^{2+}-activated protease (Tahara and Traugh 1981); (b) Ca^{2+} inhibition of eIF-2(αP) protein phosphatase; and (c) Ca^{2+} activation of the double-stranded RNA-activated translational inhibitor (DAI), but all of these possibilities were virtually ruled out (data not shown).

The work of Kuo and co-workers (Mazzei et al. 1982), made us try the effect of polymyxin B on the translational inhibition by Ca^{2+} in reticulocyte lysates. These authors reported that polymyxin B is a selective inhibitor of phospholipid and Ca^{2+}-dependent protein kinases (PL-Ca-PK). Polymyxin B at a concentration of 8 μM, while inhibiting PL-Ca-PK, was practically without effect on both cyclic AMP-dependent and cyclic GMP-dependent protein kinases. As shown in Table 4, lysates that respond poorly to the addition of hemin are stimulated by the porphyrin to a high degree when EGTA or polymyxin B (8 μM) is present. Furthermore, the strong translational inhibition promoted by Ca^{2+} is largely prevented by polymyxin B. Consistent with this result, phosphatidylserine in the presence of a small amount of diolein inhibits protein syn-

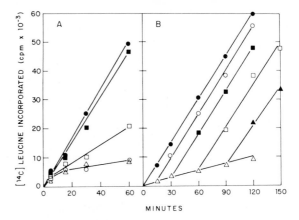

Fig. 2 A, B. Inhibition of translation by Ca^{2+} and reversal by EGTA at low incubation temperatures. **A** Incubation at 30°C. The concentrations of Ca^{2+} and EGTA, when present, were 0.5 mM and 2 mM, respectively. (●) + hemin, (○) - hemin, (△) either + hemin + Ca^{2+} with no further additions or + hemin + Ca^{2+} from the start of incubation and EGTA added after 15 or 30 min of incubation, (■) + hemin + Ca^{2+} + EGTA at the start of incubation, (□) + hemin + Ca^{2+} from the start, and EGTA added after 5 min of incubation. **B** Incubation at 20°C. (●) + hemin, (△) + hemin + Ca^{2+}, (○), (■), (□), and (▲) + hemin + Ca^{2+} from the start and EGTA added after 15, 30, 60, and 90 min of incubation, respectively. The specific radioactivity (mCi/mmol) of [^{14}C] Leucine was 54 in **A** and 342 in **B**

thesis in reticulocyte lysates supplemented with hemin (Fig. 4A and B) and this inhibition is prevented by polymyxin B (Fig. 4C). Moreover, as shown in Fig. 5, addition of phospholipid promotes the activation of HCI in hemin-supplemented reticulocyte lysates (lane 3) and this activation is prevented by either EGTA (lane 4) or polymyxin B (lane 5). The antibiotic had no effect on the activation of HCI by either hemin lack or NEM (data not shown). Phospholipid also promoted significant phosphorylation of the 90 000 mol. wt. band of HCI (data not shown).

4 Discussion

We have shown that the hemin-controlled translational inhibitor (HCI) of reticulocyte lysates can be activated in the presence of hemin, by calcium ions, and phospholipid. Although activation occurs when either Ca^{2+} or phospholipid is added alone, such activation is probably due to the presence of small amounts of endogenous Ca^{2+} and phospholipid in lysates. Our results are consistent with this idea because lysates that are poorly stimulated by hemin can be converted to highly efficient lysates by the addition of either the Ca^{2+} chelator EGTA or the phospholipid antagonist polymyxin B. Moreover, the activation of HCI by either Ca^{2+} or phospholipid can be largely prevented by either EGTA or polymyxin B (Table 4 and Fig. 5).

The Ca^{2+}-phospholipid mechanism cannot be involved in the activation of HCI due to lack of heme for neither EGTA nor polymyxin B can relieve the effect of heme

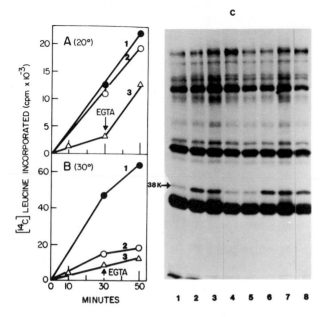

Fig. 3 A-C. Phosphorylation of the 38 000 mol. wt. (α) subunit of eIF-2 promoted by Ca^{2+} at 20°C or 30°Cis reversed by EGTA at the lower, but not at the higher temperature. Assay conditions were according to published procedures (de Haro et al. 1983). **A** and **B**, translation at 20°C and 30°C, respectively. (•) + hemin, (○) - hemin, (△) + hemin + Ca^{2+} (0.6 mM) from the start and EGTA (2 mM) added at 30 min. Aliquots for the phosphorylation assay were taken as follows: *Curves 1 and 2*, at 30 min; *Curve 3*, at 10 and 50 min; **C** Phosphorylation assay. *Tracks 1-4*, aliquots from incubations shown in **A**; 5-8 aliquots from incubations shown in **B**. *Lanes 1 and 5 (curve 1)*, with hemin; *lanes 2 and 6 (curve 2)*, without hemin; *lanes 3 and 7 (curve 3)*, with hemin and Ca^{2+} (aliquots taken at 10 min); *lanes 4 and 8 (curve 3)*, with hemin and Ca^{2+} from the start, EGTA added at 30 min (aliquots taken at 50 min)

deficiency. The biological significance, if any, of the Ca^{2+}-phospholipid mechanism remains to be determined. The Ca^{2+} and phospholipid-activated protein kinases are largely membrane-associated (Sano et al. 1983; Helfman et al. 1983) and a Ca^{2+}-phospholipid mechanism might be involved in control of protein synthesis in the cell membrane.

Table 2. Translational inhibition by Ca^{2+} is not prevented by anti-calmodulin

Additions[a]	[^{14}C] Leucine incorporation cpm x 10^{-3}		Inhibition %
	Total	Due to hemin	
None	10.1		
Hemin	36.0	25.9	
Anti-calmodulin (3 μg)	11.4		
Hemin + anti-calmodulin	32.6	21.2	
Hemin + Ca^{2+} (0.17 mM)	26.2	16.1	38
Hemin + Ca^{2+} (0.34 mM)	17.9	7.8	70
Hemin + anti-calmodulin + Ca^{2+} (0,17 mM)	24.4	13.0	39
Hemin + anti-calmodulin + Ca^{2+} (0.34 mM)	16.6	5.2	76

[a]Reticulocyte lysates were preincubated with anti-calmodulin (New England Nuclear) for 4.5 h at 0°C in conditions of translation before adding $CaCl_2$

Table 3. Calmodulin does not affect protein synthesis in reticulocyte lysates

Additions[a]	Hemin-dependent translation, cpm x 10^{-3}	Activity %
None	82.0	100
Ca^{2+}, 20 μM	76.7	93
CaM_R (200 ng)	78.0	95
CaM_R + Ca^{2+}	82.7	100
CaM_B (100 ng)	80.0	98
CaM_B + Ca^{2+}	74.9	91
CaM_S (150 ng)	83.0	100
CaM_S + Ca^{2+}	71.3	87

[a]Incorporations with and without hemin were, respectively, 98 200 and 16 200 cpm. CaM_R, calmodulin purified from reticulocyte lysates; CaM_B, calmodulin from pig brain (Boehringer); CaM_S, calmodulin from bovine brain (Sigma). All preparations of CaM were active in the phosphodiesterase stimulation assay. (not shown)

Table 4. Effect of EGTA and polymyxin B on translation

Experiment No.	Additions	[^{14}C]-Leucine incorporation cpm x 10^{-3}		Activity %
		Total	Due to hemin	
1	None	5.6		
	Hemin	14.7	9.1	100
	Hemin + EGTA (0.1 mM)	28.2	22.6	248
	Hemin + polymyxin B (8 μM)	26.2	20.6	226
	Hemin + polymyxin B (16 μM)	24.2	18.6	204
	Hemin + polymyxin B (160 μM)	22.3	16.7	183
	Hemin + Ca^{2+} (0.1 mM)	7.5	1.9	21
	Hemin + Ca^{2+} + polymyxin B (16 μM)	12.0	6.4	70
	Hemin + Ca^{2+} + polymyxin B (160 μM)	20.8	15.2	167
2	None	5.6		
	Hemin	13.6	8.0	100
	Hemin + EGTA (0.1 mM)	30.3	24.7	309
	Hemin + polymyxin B (8 μM)	27.2	21.6	270
	Hemin + polymyxin B (16 μM)	25.5	19.9	249
	Hemin + polymyxin B (160 μM)	20.3	14.7	184
	Hemin + Ca^{2+} (0.1 mM)	7.2	1.6	20
	Hemin + Ca^{2+} + polymyxin B (16 μM)	12.6	7.0	87
	Hemin + Ca^{2+} + polymyxin B (160 μM)	19.7	14.1	176

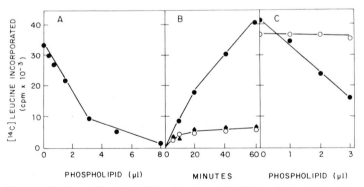

Fig. 4 A-C. Translational inhibition by phospholipid in reticulocyte lysates and reversal by polymyxin B. **A** Inhibition as a function of the concentration of phospholipid; **B** Kinetics: (•) plus hemin, (○) minus hemin (▲) plus hemin and phospholipid (4 μl); **C** Polymyxin B reversal of translational inhibition by phospholipid: (•) without, (○) with polymyxin B (50 μM). Assay conditions were according to published procedures (de Haro et al. 1983). *Curves* in **A** and **C** give the hemin-dependent translation. The phospholipid solution contained 1.5 μg of phosphatidylserine and 0.15 μg of diacylglycerol (1,3 diolein) per μl

Fig. 5. Phospholipid-promoted phosphorylation of the 38 000 mol. wt. (α) subunit of eIF-2 is prevented by either EGTA or polymyxin B. Assay conditions were as reported (de Haro et al. 1983). *Lanes: 1*, without hemin; *2*, with hemin; *3*, with hemin and phospholipid (2 μl); *4*, with hemin, phospholipid, and EGTA (5.0 mM); *5*, with hemin, phospholipid, and polymyxin B (0.8 mM)

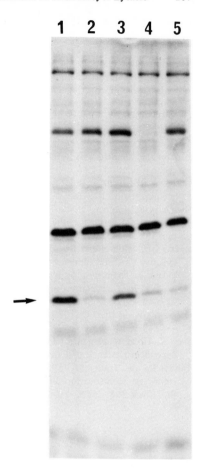

References

de Haro C, de Herreros AG and Ochoa S (1983) Activation of the heme-stabilized translational inhibitor of reticulocyte lysates by calcium ions and phospholipid. Proc Natl Acad Sci USA 80: 6843-6847

de Herreros AG, de Haro C and Ochoa S (1983) Does calcium activate the heme-controlled translational inhibitor (HCI) in reticulocyte lysates?. Fed Proc 42: 1795

Farrell PJ, Balkow K, Hunt T, Jackson RJ and Trachsel H (1977) Phosphorylation of initiation factor eIF-2 and the control of reticulocyte protein synthesis. Cell 11: 187-200

Gross M and Mendelewski J (1977) Additional evidence that the hemin-controlled translational repressor from rabbit reticulocytes is a protein kinase. Biochem Biophys Res Commun 74: 559-569

Gross M and Rabinovitz M (1972) Control of globin synthesis by hemin. Factors influencing formation of an inhibitor of globin chain initiation in reticulocyte lysates. Biochim Biophys Acta 287: 340-352

Helfman DM, Applebaum BD, Vogler WR and Kuo JF (1983) Phospholipid-sensitive Ca^{2+}-dependent protein kinase and its substrates in human neutrophils. Biochem Biophys Res Commun 111: 847-853

Hunt T, Vanderhoff G and London IM (1972) Control of globin synthesis: the role of heme. J Mol Biol 66: 471-481

Jagus R and Safer B (1981) Activity of eukaryotic initiation factor 2 is modified by processes distinct from phosphorylation. J Biol Chem 256: 1317-1323

Kramer G, Cimadevilla JM and Hardesty B (1976) Specificity of the protein kinase activity associated with the hemin-controlled repressor of rabbit reticulocytes. Proc Natl Acad Sci USA 73: 3078-3082

Levin DH, Ranu RS, Ernst V and London IM (1976) Regulation of protein synthesis in reticulocyte lysates. Phosphorylation of methionyl-tRNA$_f$ binding factor by protein kinase activity of translational inhibitor isolated from heme-deficient lysates. Proc Natl Acad Sci USA 73: 3112-3116

London IM, Clemens MJ, Ranu RS, Levin DH, Cherbas LF and Ernst V (1976) The role of hemin in the regulation of protein synthesis in erythroid cells. Fed Proc 35: 2218-2222

Mathews MB, Hunt T and Brayley A (1973) Specificity of the control of protein synthesis by haemin. Nature New Biol 243: 230-233

Maxwell CR, Kamper CS and Rabinovitz M (1971) Hemin control of globin synthesis: an assay for the inhibitor formed in the absence of hemin and some characteristics of its formation. J Mol Biol 58: 317-327

Mazzei GJ, Katoh N and Kuo JF (1982) Polymyxin B is a more selective inhibitor for phospholipid-sensitive Ca^{2+}-dependent protein kinase than for calmodulin-sensitive Ca^{2+}-dependent protein kinase. Biochem Biophys Res Commun 109: 1129-1133

Ochoa S (1983) Regulation of protein synthesis initiation in eukaryotes. Arch Biochem Biophys 223: 325-349

Ochoa S and de Haro C (1979) Regulation of protein synthesis in eukaryotes. Ann Rev Biochem 48: 549-580

Sano K, Takai Y, Yamanishi J and Nishizuka Y (1983) A role of calcium-activated phospholipid-dependent protein kinase in human platelet activation. J Biol Chem 258: 2010-2013

Tahara SM and Traugh JA (1981) Cyclic nucleotide-independent protein kinases from rabbit reticulocytes: identification and characterization of a protein kinase activated by proteolysis. J Biol Chem 256: 11558-11564

High pO$_2$-Promoted Activation of a Translational Inhibitor in Rabbit Reticulocytes: Role of the Oxidation State of Sulfhydryl Groups

B. Kan[1], G. Kanigür[1], D. Tiryaki[1], N. Gökhan[2], and E. Bermek[1]

1 Introduction

Different experimental conditions in reticulocytes or their lysates cause the activation of kinases specific for the α-subunit of the eukaryotic initiation factor 2 (eIF-2) with the subsequent inhibition of protein synthesis (for reviews see Revel and Groner 1978; Safer and Anderson 1978; Ochoa and de Haro 1979; Ochoa 1983 and related papers of this volume). With few expections, however, it is difficult to correlate these experimental conditions to those prevailing in the erythroid cell. Considering that the red blood cells due to their functions represent cellular elements particularly exposed to O$_2$ effects, we asked whether a control mechanism of globin synthesis dependent on pO$_2$ exists. The results revealed a reciprocal relationship between the activity of protein synthesis and pO$_2$, and also the activation of an inhibitor of protein synthesis (Almiş-Kanigür et al. 1982). Moreover, the high pO$_2$ (hpO$_2$)-mediated inhibition appeared to share common features with other cases of translational inhibition in reticulocytes and, in particular, with the inhibition caused by oxidized glutathione (GSSG). The data presented below indicate that both hpO$_2$ and GSSG can activate a heat-stable inhibitor of $\sim$23000 in reticulocyte lysates.

2 Results

As shown in Fig. 1, the incubation of reticulocytes under hpO$_2$ results in a decrease of their activity in protein synthesis. In comparison to the control group kept under normal humidified atmospheric air, that is, pO$_2$ of 19 kPa, the reticulocytes incubated in the presence of pO$_2$ of 37 kPa reveal in their activity an inhibition of 70%-80%. On the other hand, the reticulocytes incubated at pO$_2$ of about 13 kPa display a protein synthetic activity more than twofold higher than that of the control reticulocytes.

---

[1] Department of Biophysics, Istanbul Medical Faculty, Istanbul University, Çapa, Istanbul, Turkey
[2] Department of Physiology, Istanbul Medical Faculty, Istanbul University, Çapa, Istanbul, Turkey

---

Mechanisms of Protein Synthesis
ed. by Bermek
© Springer-Verlag Berlin Heidelberg 1984

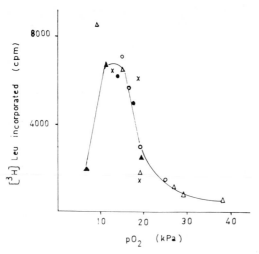

Fig. 1. The effect of pO_2 upon protein synthesis in reticulocytes. The curve is a compilation of five different assays indicated by the different symbols. The determined pCO_2 and pH values for the samples were $0.81 (\mp 0.29)$ kPa and $6.95 (\mp 0.2)$, respectively

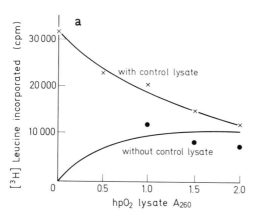

Fig. 2. The effect of hpO_2 upon protein synthesis in the lysates from rabbit reticulocytes. Activation of an inhibitor of protein synthesis in the lysate under hpO_2. Control and hpO_2 lysates were prepared as described (Almiş-Kanigür et el.1982). Indicated amounts of hpO_2 lysates were added to 1.0 A_{260} unit of control lysate and assayed in endogeneous synthesis. [^{3}H] leucine incorporated into hot CCl_3COOH-precipitable material was determined in 40 μl aliquots from reaction mixtures of 50 μl

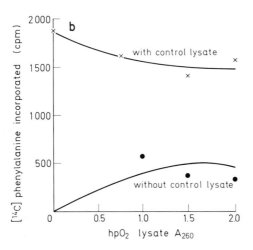

Fig. 3. The effect of hpO_2 upon protein synthetic activity of rabbit reticulocytes as assayed in polyphenylalanine synthesis. Absence of a hpO_2-promoted inhibitory activity specific for polyphenylalanine synthesis. Control and hpO_2 lysates were prepared as described (Almiş-Kanigür et al. 1982). Indicated amounts of hpO_2 lysates were added to 1 A_{260} unit of control lysates and assayed in polyphenylalanine synthesis. [^{14}C]phenylalanine incorporated into hot CCl_3COOH-precipitable material was determined in 40 μl aliquots from reaction mixtures of 50 μl

Thus, protein synthetic activity appears to be highest under pO$_2$ values prevailing in the arterial blood. The normal atmospheric pO$_2$, on the other hand, proves to be too high for reticulocyte protein synthesis.

As shown in Figs. 2 and 3, the loss of protein synthetic activity promoted by hpO$_2$ is observed also in the lysates prepared from them both in leucine incorporation and polyphenylalanine synthesis. Moreover, an inhibition of leucine incorporation becomes apparent in proportion to the amount of the hpO$_2$ lysate added: at 1 to 2 A$_{280}$ ratio of the control to the hpO$_2$ lysate, protein synthesis is depressed to the level of that in the hpO$_2$ lysate (Fig. 2). However, the hpO$_2$ lysate seems to have negligible effect on the control activity in polyphenylalanine synthesis (Fig. 3). In addition, the analysis by sucrose density gradient centrifugation shows no difference in polysomal profiles from the control vs hpO$_2$ reticulocytes. In particular, there is no sign of a polysomal degradation in the hpO$_2$ lysate (data not shown). The results from leucine incorporation and polyphenylalanine synthesis together with those from sucrose density gradient centrifugation imply that the hpO$_2$ effects on protein synthesis are complex: besides mediating the activation of an inhibitor of endogeneous protein synthesis, hpO$_2$ seems to cause the inactivation of some components involved in chain elongation. We tried to determine the nature of the inhibitor and its relationship to other inhibitors of translation.

The rate of protein synthesis in the hpO$_2$ lysate is maintained during the first 3-5 min. at the control level, thereafter, leveling off to give biphasic kinetics (Fig. 4). The inhibition in the high pO$_2$ lysate is potentiated by ATP (Fig. 5) and overcome by high concentrations (2 mM) of cAMP (Fig. 6). As shown in Table 1, the activation of hpO$_2$ inhibitor can also be prevented by the addition of glucose-6-phosphate (G6P) to lysate prior to treatment with hpO$_2$. The biphasic kinetics of inhibition, its potentiation by

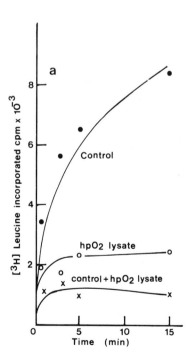

Fig. 4. Activation of hpO$_2$I in reticulocyte lysates under high pO$_2$. Experimental procedure was as described (Almiş-Kanigür et al. 1982)

Table 1. Effect of G6P on the activation of hpO_2I^a

System	[³H] leucine incorporated (cpm)
Control lysate	944
hpO_2 lysate	97
Control lysate + Control lysate	1335
Control lysate + hpO_2 lysate	180
Control lysate + Control lysate + G6P	1283
Control lysate + hpO_2 lysate + G6P	1485

[a]Experimental procedure was as described (Almiş-Kanigür et al. 1982).

ATP, and its prevention by high concentrations of cAMP or by G6P implicates a relationship of hpO_2 I to hemin-controlled repressor (HCR) and to glutathione disulfide-activated inhibitor (GSSGI) (Kosower et al. 1972; Giloh (Freudenberg) et al. 1975; Ernst et al. 1978; Lenz et al. 1978; Wu 1981).

High pO_2 I activity is found in the ribosomal salt wash fraction from the hpO_2 lysates, but not in S-100 (data not shown). When the wash fraction from the hpO_2 lysate is added to the control lysate, it decreases leucine incorporation nearly to the level obtained with edeine (Table 2). On the other hand, edeine plus the wash fraction from hpO_2 lysate depresses leucine incorporation slightly below the level obtained with the inhibitors used separately. This finding suggestes that the inhibitory effect of hpO_2 I on endogeneous protein synthesis is due to the inhibition of chain initiation.

The inhibitory activity can be partially purified from the S-100 wash fraction of the hpO_2 lysate (hpO_2-S-100W) by precipitation between 0 to 50% ammonium sulfate

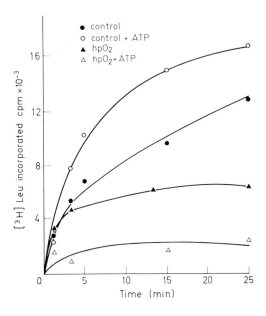

Fig. 5. Effect of ATP upon the activation of hpO_2I in reticulocyte lysates. Experimental procedure was as described (Almiş-Kanigür et al. 1982)

Fig. 6. Effect of cAMP upon the activation of hpO$_2$I in reticulocyte lysates. Experimental procedure was as described (Almiş-Kanigür et al. 1982)

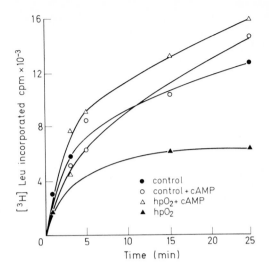

saturation and chromatography of this fraction on Sephadex G-100 (Fig. 7). The hpO$_2$I-activity is eluted slightly behind the main A$_{280}$ peak on a Sephadex G-100 column at a region corresponding to a mol. wt. of ~23000. Additionally, two minor peaks of inhibitory activity are detected, one located behind the void volume and the other in the low molecular weight region corresponding to column volume. Considering the close relationship between the effects of pO$_2$ and the oxidation of sulfhydryl groups, we investigated the relationship between hpO$_2$I and GSSGI. The S-100W fraction from control lysates was treated with GSSG as described (Ernst et al. 1978), passed through Sephadex G-100 column, and the column fractions were assayed for inhibitory activity. The GSSGI activity is primarily found, like the hpO$_2$I, in the 23000 region of the eluate. The inhibitory activity eluting behind the void volume is observed also in this case. The inhibitor can also be activated by direct treatment of the fractions of the control eluate corresponding to 23000 region with either hpO$_2$ or GSSG. The presence of ATP seems to be essential for the hpO$_2$-dependent activation of the fractionated

Table 2. Comparison of the effects of edeine and of hpO$_2$I on leucine incorporation[a]

System[a]	[^{3}H] leucine incorporated (cpm)
Control lysate	4523
Control lysate + edeine	2164
hpO$_2$ lysate	1540
hpO$_2$ lysate + edeine	1059
Control lysate + hpO$_2$-S-100(W)	2270
Control lysate + hpO$_2$-S-100(W) + edeine	1724

[a]Control and hpO$_2$ lysates prepared as described (Almiş-Kanigür et al., 1982) were assayed for protein synthesis in the presence and absence of 1.5 μM edeine. Wherever specified, 50 μg of hpO$_2$-S-100(W) was added. [^{3}H]leucine incorporated into hot CCl3COOH-precipitable material was determined

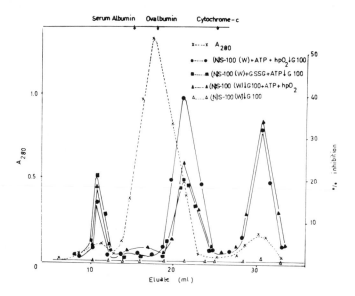

Fig. 7. Chromatography of hpO$_2$I and of GSSGI on Sephadex G-100. S-100(W) proteins were treated either with hpO$_2$ or with GSSG (Ernst et al. 1978) and gel-filtered on Sephadex G-100. Alternatively, the S-100(W) fraction without treatment(=(N)S-100 (W)) was directly applied to G-100 column. Whenever indicated, G-100 fractions of (N)S-100(W) were subsequently treated with hpO$_2$ and GSSG. Aliquots from G-100 fractions were assayed in protein synthesis for inhibitory activity. The M$_r$ of the inhibitor eluting behind the main A$_{280}$ peak was estimated to be ~23000. For details of the experimental procedure sel. (Almiş-Kanigür et al. 1982 and Kanigür et al. 1983)

inhibitor (Table 3) and, as is the case with the lysate and S-100W, G6P or high concentrations of cAMP are found to prevent its activation. The ~23000 fraction from the pO$_2$-treated S-100W, like the pressure activated inhibitor, PAI, (Henderson et al. 1979) is stable and activable during a 5 min incubation of 80°C (Table 3).

The data above suggested a parallelism between the hpO$_2$-and/or GSSG-activated inhibitor of ~23000 and a sulfhydryl-protein of similar molecular weight (Jackson et al. 1983). The inactivation of the Sephadex G-25 gel-filtered reticulocyte lysates in protein syntesis seemed to be accompanied by an oxidation of the SH-group(s) of this latter protein. We decided, therefore, to investigate the relationship of the activation of hpO$_2$I to this 24000 dalton SH-protein. For this purpose, the ~23000 region of the G-100 fractions of the control, either directly or after treatment with hpO$_2$ and the corresponding G-100 fractions from the pO$_2$-treated ribosomal salt wash were incubated with [^{14}C]N-ethylmaleimide. The SDS polyacrylamide gel electrophoresis of these three samples and the subsequent determination of the radioactivity in the gels sliced showed a high ^{14}C radioactivity peak and a hpO$_2$-dependent loss of the label in the region of 23000 in parallel to inhibitor activation (Fig. 8). Thus, a role of the oxidation state of this protein ist implicated in the activation of hpO$_2$I and possibly of GSSGI.

Table 3 Activation of the ~23 000-M$_r$ inhibitor of protein synthesis by hpO$_2$ or GSSG (2.5 mM); effects of G6P, ATP, cAMP or of heat treatment upon activation

System	[^{3}H] leucine incorporated cpm/1 A$_{260}$ unit lysate
	Experiment I
Control lysate	18993
+ ~23 000-M$_r$ fraction 5 min 37°C (+ATP)	21283
+ ~23 000-M$_r$ fraction 5 min 37°C (+ATP), hpO$_2$	12330
	Experiment II
Control lysate	7588
+ ~23 000-M$_r$ fraction 5 min 37°C	6667
+ ~23 000-M$_r$ fraction 5 min 37°C hpO$_2$	5972
+ ~23 000-M$_r$ fraction 5 min 37°C (+ATP), hpO$_2$	1880
+ ~23 000-M$_r$ fraction 5 min 37°C (+G6P), hpO$_2$	6663
+ ~23 000-M$_r$ fraction 5 min 37°C (+cAMP), hpO$_2$	8783
+ ~23 000-M$_r$ fraction 5 min 37°C (+GSSG)	4272
+ ~23 000-M$_r$ fraction 5 min 37°C (+GSSG+ATP)	2957
+ ~23 000-M$_r$ fraction 5 min 37°C (+GSSG+G6P)	7428
+ ~23 000-M$_r$ fraction 5 min 80°C	5473
	Experiment III
Control lysate	13703
+ ~23 000-M$_r$ fraction from hpO$_2$-S-100(W)	7400
+ ~23 000-M$_r$ fraction from hpO$_2$ 5 min 80°C	7908

Experimental procedure was as described (Kanigür et al. 1983)

3 Discussion

The results implicate that globin chain synthesis can be controlled by the ligand concentration (pO$_2$) and that a reciprocal relationship exists between protein synthetic activity and pO$_2$. The elevated protein syntetic activity of reticulocytes under low (10-15 kPa) pO$_2$ might reflect the actual physiological in vivo conditions in which the red blood cells function.

High pO$_2$-promoted inhibition of protein synthesis appears to be complex: besides invoking the activation of an inhibitor of protein synthesis, it results in the inactivation of some components involved in polypeptide chain elongation. The hpO$_2$-promoted inhibition of elongation which has not been the primary object of this study seems to be due to the oxidation of some ribosomal sulfhydryl groups and of those of elongation factor 2.

The inhibitor activated under hpO$_2$ reveals, in respect to its mechanism of action, similarities with other inhibitors of protein synthesis, i. e., it causes biphasic kinetics of inhibition, its activation is promoted by ATP and prevented by high concentrations of cAMP and of G6P. Furthermore, the data obtained with edeine implicates chain initiation to be its site of attack, like that of the other inhibitors. However, its apparent mol. wt. of ~ 23 000 attests to an important distinction from HCR or from double-stranded

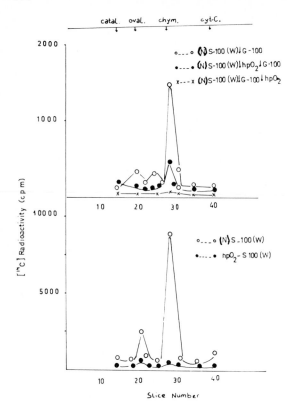

Fig. 8. High pO_2-promoted oxidation of sulfhydryl groups(s) in S-100(W) and in G-100 fractions of ~23000 as assayed by electrophoresis on sodium dodecyl sulfate polyacrylamide gels after reaction with [^{14}C]N-etylmaleimide. *Upper panel*: (○--○), control G-100 fraction of ~23000; (x--x), control fraction of ~23000 subsequently treated with hpO_2; (●--●), G-100 fraction of ~23000 from hpO_2-S-100(W). *Lower panel*: (○--○), control S-100W); (●--●), hpO_2-S-100(W). Standards from *left* to *right, catal,* catalase; *oval.,* ovalbumin; *chym.,* chymotrypsinogen; *cyt. C.,* cytochrome c. Details of the experimental procedure, see (Kanigür et al. 1983)

RNA activated inhibitor (dsRNAI). On the other hand, hpO_2I and GSSGI both elute from G-100 column at same region. This finding together with the well-known cellular functions of GSH suggests that the effects of hpO_2 and of GSSG might indeed occur through the same mechanism, that is, through the activation of an inhibitor protein of ~23000. The similarities in chromatographic behavior and the heat stability of this inhibitory protein indicate also a close relationship to the heat stable component of the pressure activated inhibitor (PAI) (Henderson et al. 1979). Furthermore, the data attest to a relationship of hpO_2I (GSSGI) to the sulfhydryl protein of similar molecular weight, possibly involved in its oxidized state in the inactivation of the G-25 gel-filtered reticulocyte lysates. Thus, in all the mentioned cases the oxidation of a sulfhydryl protein of ~23000 is implicated in the inactivation of the lysates. The relationship of this protein to HCR and/or to the kinase(s) specific for the α-subunit of eIF-2 remains to be established. It is likely that it plays a mediating role between the oxidation state of the cell and the cascade system resulting at the end in the activation of the eIF-2α kinase. According to our data, the activation of this protein seems to be under additional control of ATP, G6P, or cAMP. Hence, the presence of such a protein may warrant a certain differentiation in the response of the cell to oxidation effects and provide the dependence of oxidation-promoted activation of eIF-2α kinase on the level of ATP and G6P in the cell.

Acknowledgement. This investigation was supported by a grant (TAG-495) from the Scientific and Technical Research Council of Turkey (TÜBITAK).

References

Almiş-Kanigür G, Kan B, Kospançali S, Bermek E (1982) A Translational inhibitor activated in rabbit reticulocyte lysates under high pO$_2$. FEBS Lett 145: 143-146

Ernst V, Levin DH, London IM (1978) Inhibition of protein synthesis initiation by oxidized glutathione: Activation of a protein kinase that phosphorylates the α-subunit of eukaryotic initiation factor 2. Proc. Natl Acad Sci USA 75: 4110-4112

Giloh (Freudenberg) H, Mager J (1975) Inhibition of peptide chain initiation in lysates from ATP-depleted cells, I. Stages in the evolution of the lesions and reversal by thiol compounds, cyclic AMP or purine derivatives and phosphorylated sugars. Biochim Biophys Acta 414: 293-308

Giloh (Freudenberg) H, Schochot L, Mager J (1975) Inhibition of peptide chain initiation in lysates from ATP-depleted cells. II. Studies on the mechanism of the lesion and its relation to similar alterations caused by oxidized glutathione and hemin deprivation. Biochim Biophys Acta 414: 309-320

Henderson HB, Miller AH, Hardesty B (1979) Multistep regulatory system for activation of a cyclic AMP-independent eukaryotic initiation factor-2 kinase. Proc Natl Acad Sci USA 76: 2605-2609

Jackson RJ, Herbert P, Campbell EA, Hunt T (1983) The roles of sugar phosphates and thiol-reducing systems in the control of reticulocyte protein synthesis. Eur J Biochem 131: 313-324

Kanigür G, Kan B, Tiryaki D, Bermek E (1983) High pO$_2$-activated inhibitor of protein synthesis in rabbit reticulocytes: Its relationship to glutathione disulfide-induced inhibitor and to a ~ 23 000-M$_r$ sulfhydryl protein. Biochem Biophys Res Commun 117: 135-140

Kosower NS, Vanderhoff GA, Kosower EM (1972) Glutathione VIII: The effects of glutathione disulfide on initiation of protein synthesis. Biochim Biophys Acta 272: 623-637

Lenz JR, Chatterjee GE, Maroney PA, Baglioni C (1978) Phosphorylated sugars stimulate protein synthesis and Met-tRNA$_f$ binding activity in extracts of mammalian cells, Biochemistry 17: 80-87

Ochoa S (1983) Regulation of protein synthesis initiation in eukaryotes. Arch Biochem Biophys 223: 325-349

Ochoa S, deHaro C (1979) Regulation of protein synthesis in eukaryotes. Annu Rev Biochem 48: 549-580

Revel M, Groner Y (1978) Post-transcriptional and translational controls of gene expression in eukaryotes. Annu Rev Biochem 47: 1079-1126

Safer B, Anderson WF (1978) The molecular mechanism of hemoglobin synthesis and its regulation in the reticulocyte, Crit Rev Biochem 5: 261-289

Wu JM (1981) Effects of phosphorylated sugars on the formation of an inhibitor of protein synthesis activated by oxidized glutathione. FEBS Lett 133: 107-111

Eukaryotic Chain Elongation

Guanine Nucleotide Binding to Eukaryotic Elongation Factor 2: A Possible Control Mechanism of Eukaryotic Chain Elongation

E. BERMEK, R. NURTEN, N. BILGIN-AKTAR, Ü.G. ODABAŞ, and O. Z. SAYHAN[1]

1 Introduction

Chain elongation in eukaryotes appears to occur in a principally similar manner as in prokaryotes with three protein factors involved in its different steps (Miller and Weissbach 1977; Bermek 1978; Clark 1980; Bielka 1982). One of these three factors, elongation factor 2 (EF-2), which participates in the translocation on the eukaryotic ribosome has properties mostly in common with its prokaryotic counterpart elongation factor G (EF-G). Nevertheless, some distinct properties do distinguish EF-2 from EF-G. First, EF-2 represents the sole site of attack of diphtheria toxin and of the exotoxin from *P. aeruginosa* (Collier 1967; Honjo et al. 1968; Iglewski and Kabat 1975). Both toxins inactivate EF-2 through adenosine diphosphate (ADP)-ribosylation of its diphthamide moiety which seems to be unique for this protein (van Hess et al. 1980a; van Hess et al. 1980b). EF-2 displays, moreover, RNA-binding properties like some other protein factors involved in the translation of the eukaryotic message (Spirin and Ovchinnikov 1983). Furthermore, EF-2 seems to have distinctions from EF-G also in respect to its interactions with guanine nucleotides. Namely it can form stable complexes with guanine nucleotides in contrast to EF-G which has only transient interactions with them shown by indirect ways (Raeburn et al. 1968; Bodley et al. 1970; Montanaro et al. 1971; Bermek and Matthaei 1971; Chuang and Weissbach 1972; Girshovich et al. 1976; Arai et al 1975). The results reported below implicate the involvement of the direct interactions of GDP (and of ADP) with EF-2 in the modulation of its activity during polypeptide chain elongation.

2 Results

Investigations carried out with purified (rat or hog liver) EF-2 preparations have shown a higher affinity of the factor for GDP in the formation of the binary complex than for GTP (Mizumoto et al. 1974; Henriksen et al. 1975a; Nurten and Bermek 1980): with rat liver EF-2, the apparant K_d values for GDP, GTP, and GuoPP(CH_2)P have been found to be 0.5, 2,7, and 24.0 μM, respectively (Table 1); these data are in good agree-

---

[1] Department of Biophysics, Istanbul Medical Faculty, Istanbul University, Çapa, Istanbul, Turkey

Mechanisms of Protein Synthesis
ed. by Bermek
© Springer-Verlag Berlin Heidelberg 1984

Table 1. Binding of guanine nucleotides to EF-2

Components present in binding systems	K_d of binding reaction (μM)	Number of binding sites (n)	Previously published data K_d (μM)	
EF-2, GDP	2.68	0.80	14.0^c	Mizumoto, Iwasaki, Kaziro (1974)
			2.0^a	Henriksen, Robinson, Maxwell (1975a)
EF-2, GDP	0.50	0.77	0.32^b	Chuang, Weissbach (1972)
			0.41^c	Mizumoto, Iwasaki, Kaziro (1974)
			0.40^a	Henriksen, Robinson, Maxwell (1975a)
EF-2, GuoPP(CH_2)P	24.0	0.60	20.0^a	Henriksen, Robinson, Maxwell (1975a)
EF-2, GDP$_{ox}$	38.0	0.75		
EF-2, GDP$_{ox\text{-}red}$	26.5	1.10		
EF-2, GTP-Mg^{+2}	2.77	0.70		
EF-2, GDP-Mg^{+2}	1.57	0.80		

[a] Determined by equilibrium dialysis
[b] Determined by filter binding
[c] Determined by column chromatography
Experimental procedure was as described (Nurten and Bermek 1980)

ment with those obtained also with rat liver EF-2 by Henriksen et al. (1975a). Thus, the affinity of rat liver EF-2 for GDP is about five times higher than that for GTP, but this difference decreases in the presence of Mg^{2+}, with the apparent K_d values of 1.6 and 2.8 μM for GDP and GTP.

In view of the ratio of the intracellular concentrations of nucleotide diphosphates to those of nucleotide triphosphates and of the ratio of the apparant K_d values for GDP and for GTP in the binding to EF-2, it might be assumed that EF-2 can be directly regenerated by GTP from the hypothetical EF-2. GDP complex released from the post-translocational ribosomes:

$$EF\text{-}2. GDP + GTP \rightleftharpoons EF\text{-}2. GTP + GDP \qquad (1)$$
$$EF\text{-}2. GTP + ribosome \rightleftharpoons EF\text{-}2. GTP. ribosome \qquad (2)$$

Table 2. Binding of GDP and GuoPP(CH_2)P to EF-2. ribosome

Guanine nucleotide bound	K_d (μM)	Binding sites/EF-2 (n)
GDP	$9.3 (26.0)^a$	0.73
GuoPP(CH_2)P.	0.26	0.44

[a]Henriksen, Robinson, Maxwell 1975a); determined by equilibrium dialysis
Experimental procedure was as described (Nurten and Bermek 1980)

Table 3. Binding of ADPR-EF-2 to ribosomes

ADPR-EF-2 bound to ribosome	K_d (μM)	Binding sites/ ribosome (n)
− GuoPP(CH$_2$)P	0.18	1.25
+ GuoPP(CH$_2$)P	0.05	1.25

Experimental procedure was as described (Nurten and Bermek 1980)

The affinities of EF-2 for GDP vs GuoPP(CH$_2$)P, as a nonhydrolizable representative for GTP, are inversed in the presence of ribosomes, the apparant K_d values being in this case 0.26 μM for GuoPP(CH$_2$)P and 9.3 μM for GDP, respectively (Table 2).

Moreover, ADPR(EF-2), which has been used as a radioactively labeled EF-2 derivative, displays by itself a relatively high affinity for ribosomes which is further increased in the presence of GuoPP(CH$_2$)P with the apparant K_d values found as 0.18 and 0.05 μM, respectively (Table 3).

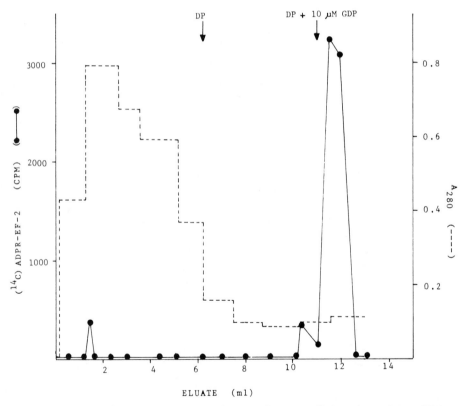

Fig. 1. Purification of EF-2 by chromatography on GDP-agarose affinity column. 3.5 mg EF-2 proteins after DE-32 step (Nurten and Bermek 1980) were applied to GDP-agarose column (0.5 x 2 cm) equilibrated with dialysis buffer (50 mM Tris-HCl pH 7.4 0.1 mM K-EDTA, 7 mM 2-mercaptoethanol). Column was washed with dialysis buffer and EF-2 was assayed by ADP-ribosylation eluted with same buffer containing 10 μM GDP

Fig. 2 a, b. SDS-polyacrylamide gel electrophoresis of EF-2 proteins **a** before and **b** after chromatography on GDP-agarose affinity column

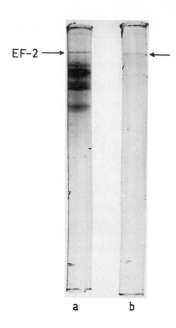

EF-2 ⟶

a   b

These findings suggest that recycling of EF-2 (and the regeneration of EF-2 from the EF-2.GDP intermediate released from the ribosome) and the subsequent formation of the ternary complex might rather occur either directly from its components:

$$\text{EF-2.GDP} + \text{ribosome} + \text{GTP} \rightleftharpoons \text{EF-2.GTP.ribosome} + \text{GDP} \qquad (3)$$

or, provided the affinities of EF-2 and of ADPR-EF-2 for the ribosome are similar, through an EF-2.ribosome intermediate:

$$\text{EF-2.GDP} + \text{ribosome} \rightleftharpoons {} + \text{EF-2. ribosome} + \text{GDP} \qquad (4)$$
$$\text{EF-2.ribosome} + \text{GTP} \rightleftharpoons \text{EF-2.GTP.ribosome} \qquad (5)$$

The latter mechanism seems to be consistent with the findings of Henriksen et al. (1975b) who have reported an inhibitory effect of ribosomes on the binding of GDP to EF-2. Results supporting their data have been obtained also using GDP-agarose affinity column. EF-2 or ADPR-EF-2 gets specifically bound to such a column and can be, thereafter, washed with GDP (Fig. 1). An effective purification of the factor can be achieved by this approach as can be seen by polyacrylamide gel electrophoretic analysis of the fractions obtained by chromatography of a crude EF-2 containing protein fraction on a GDP-agarose column (Fig. 2). Column-bound EF-2 can be specifically eluted by increasing GDP concentrations $\geqslant 1 \ \mu M$ GDP in the elution buffer. Alternatively, consistent with the results of Henriksen et al. (1975b), EF-2 seems to be washable from GDP-agarose columns with ribosomes (data not shown).

If the ternary complex is formed either directly from its components or via an EF-2.ribosome complex, EF-2.GTP complex would not necessarily be an intermediate in the formation of EF-2.GTP.ribosome complex. We carried out experiments with MalNEt and TNBS to characterize this interactions further. MalNEt has been previously shown to abolish the activity of EF-2 in polypeptide chain elongation and to

Table 4. Effects of different reagents on the quantitative binding of guanine nucleotides to EF-2

System	K_d (μM)	Binding site/EF-2 (n)	
EF-2, GDP	0.54	0.67	
EF-2, GDP + MalNEt	1.2	0.6	
EF-2, GDP – 2-mercaptoethanol	3.3	1.0	
EF-2, GDP + TNBS	–	–	(No binding)
EF-2, GDP + DMA	–	–	(No binding)
EF-2, GDP + DMS	–	–	(No binding)
EF-2, GDP + 2,3-butandione	1.2	1.0	
EF-2, GTP	2.0	0.76	
EF-2, GTP + MalNEt	2.0	0.6	
EF-2, GTP + TNBS	–	–	(No binding)
EF-2, GTP + DMS	–	–	(No binding)
EF-2, GTP + 2,3-butandione	1.93	0.8	

Experimental procedure was as described (Nurten et al 1983)

inhibit the formation of EF-2.GTP. ribosome complex. However, aside from elevating the apparant K_d value for GDP binding about twofold, MalNEt hardly affects the direct interactions between guanine nucleotides and EF-2 (Table 4). On the other hand, amino group specific (bifunctional) imidates or TNBS abolish these interactions completely. The treatment of EF-2 with MalNEt vs TNBS has, moreover, inverse effects on the formation of the ternary complex involving EF-2.GuoPP(CH$_2$)P. ribosome vs EF-2.GDP. ribosome (Table 5): with the ternary complex involving GuoPP(CH$_2$)P, TNBS seems to increase to some extent the apparant K_d (from 0.07 μM to 0.17 μM), whereas MalNEt reduces the complex formation to a negligible level (n= 0.01). On the other hand, the effects of these reagents on EF-2.GDP.ribosome complex formation are exactly opposite. TNBS treatment of EF-2 abolishes the formation of this complex, but treatment with MalNEt hardly affects it, the apparent K_d remaining 1.2 μM.

Table 5. Effect of preincubation of EF-2 with MalNEt or TNBS on the quantitative (EF-2-promoted) binding of GuoPP(CH$_2$)P. or GDP to ribosomes[a]

Additions		K_d (μM)	Binding sites/ ribosome (n)	
1. Incubation	2. Incubation			
EF-2	Ribosome + GDP	1.55 (9.3)	0.80	
EF-2	Ribosome + GuoPP(CH$_2$)P.	0.07 (0.26)	0.55	
EF-2 + MalNEt	Ribosome + GDP	1.2	0.56	
EF-2 + MalNEt	Ribosome + GuoPP(CH$_2$)P.	0.03	0.01	
EF-2 + TNBS	Ribosome + GDP	–	–	(No binding)
EF-2 + TNBS	Ribosome + GuoPP(CH$_2$)P.	0.17	0.66	

[a]The EF-2 concentration was kept at 0.15 μM and the ribosome concentration at 0.3 μM. The GuoPP(CH$_2$)P. and GDP concentration were varied between 5 nM and 75 nM, 0.5 μM, and 2.5 μM, respectively. The ribosomal complexes were isolated by centrifugation and the data evaluated as described (Nurten and Bermek 1980). The values are means of two to four separate determinations evaluated by the least squares method. The K_d values given in parantheses correspond to values determined in another set of experiments by equilibrium dialysis (see Table 2) (Nurten and Bermek 1980)

Table 6. Effect of preincubation of ADPR-EF-2 with MalNEt on the quantitative binding of ADPR-EF-2 to ribosomes[a]

Additions		K_d (μM)	Binding sites/
1. Incubation	2. Incubation		ribosome (n)
ADPR-EF-2	Ribosome	0.25 (0.18)	1.2
ADPR-EF-2	Ribosome + GDP	0.25	1.05
ADPR-EF-2	Ribosome + GuoPP(CH$_2$)P.	0.08 (0.05)	1.2
ADPR-EF-2 + MalNEt	Ribosome	0.99	0.82
ADPR-EF-2 + MalNEt	Ribosome + GDP	1.12	0.6
ADPR-EF-2 + MalNEt	Ribosome + GuoPP(CH$_2$)P.	1.01	1.0

[a]ADPR-EF-2 concentration was varied between 16.5 and 132 nM. The ribosome concentration was 0.3 μM. The concentration of guanine nucleotides, whenever added were, 20 μM. The ribosomal complex were isolated by centrifugation and the data evaluated as described (Nurten and Bermek 1980). The values are means of four separate determinations evaluated by the least squares method. The K_d values determined also by sedimentation in another set of experiments for the binding of ADPR-EF-2 to ribosomes are given in parentheses

MalNEt which specifically inhibits the EF-2.GuoPP(CH$_2$)P. ribosome complex formation does not abolish the direct interaction between ADPR-EF-2 and ribosomes (Table 6). Nevertheless, it elevates the apparant K_d value for the modified factor in this interaction from 0.25 to 0.99 μM and particularly abolishes the GuoPP(CH$_2$)P-promoted

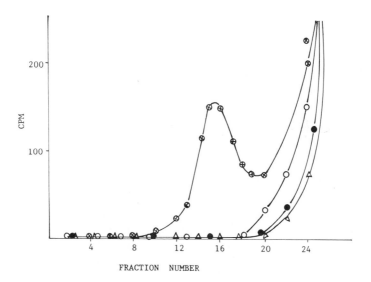

FRACTION NUMBER

Fig. 3. Effect of treatment of EF-2 with TNBS on the formation of the linkage between EF-2 and oxidized GuoPP(CH$_2$)P. EF-2 was preincubated with or without 1 mM TNBS for 5 min at 4°C. After addition of 0.3 mM (^{3}H)GuoPP(CH$_2$)P or (^{3}H)GuoPP(CH$_2$)P$_{ox}$, incubation was continued for 5 min at 37°C. Reaction mixtures containing EF-2 and (^{3}H)GuoPP(CH$_2$)P (o--o), EF-2 + TNBS, and (^{3}H)GuoPP(CH$_2$)P (●--●), EF-2 and (^{3}H)GuoPP(CH$_2$)P$_{ox}$ (⊕—⊕) or EF-2 + TNBS and (^{3}H)GuoPP(CH$_2$)P$_{ox}$ (△--△) were applied to a Sephadex G-50 column (1 x 25 cm) equilibrated with dialysis buffer. Details of the experimental procedure was as described (Nurten et al 1983)

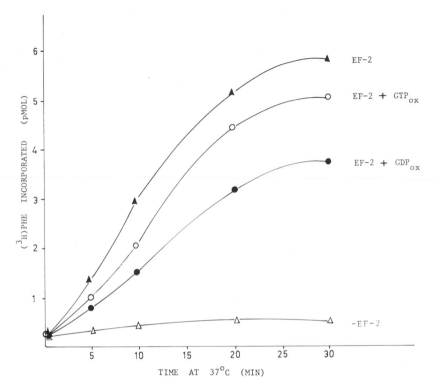

Fig. 4. Effect of treatment of EF-2 with oxidized guanine nucleotides on polyPhe synthesis. EF-2 was preincubated with or without 1 mM GTP$_{ox}$ or GDP$_{ox}$ for 5 min at 20°C and, thereafter, added to reaction mixtures for polyPhe synthesis containing 0.5 mM GTP. PolyPhe synthesis was assayed for 10 min at 37°C under conditions as described (Bermek 1976). Final concentration of GTP$_{ox}$ or GDP$_{ox}$ in reaction mixtures was 0.25 mM. (EF-2 dialysed after treatment with GDP$_{ox}$ displayed similarly reduced activity).

increase in the affinity of ADPR-EF-2 for the ribosome. GDP affects the ribosomal interactions of neither MalNet-treated nor untreated ADPR-EF-2.

The direct interactions of EF-2 with guanine nucleotides can be also investigated by using NaIO$_4$-oxidized guanine nucleotides. Normally, EF-2.GTP complex can not be detected when assayed by filtration through nitrocellulose. However, incubation of EF-2 with oxidized GTP seems to result in the formation of a product stable enough to be recovered by nitrocellulose filter technique, by precipitation in 5% CCl$_3$COOH or by SDS-polyacrylamide gel electrophoresis, and containing oxidized nucleotide and EF-2 in stoichiometric amounts (Nurten and Bermek 1980). This linkage which is likely due to an alteration of an amino group by formation of a Schiff base has not been observed in the case of EF-G. The presence of an amino group involved in direct interactions with guanine nucleotides is also implicated by the finding that TNBS abolishes this linkage in addition to the interactions between guanine nucleotides and EF-2 (Fig. 3). The linkage of oxidized nucleotides to EF-2 (that is, the blockage of its site involved in direct interactions with guanine nucleotides by oxidized nucleotides) does not, however, abolish its activity in polyphenylalanine (polyPhe) synthesis (Fig 4).

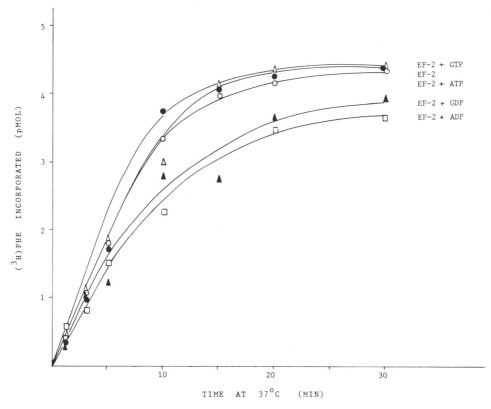

Fig. 5. Effect of treatment of EF-2 with purine nucleotides on polyPhe synthesis. EF-2 preincubated for 5 min at 20°C with or without 1 m*M* GTP, GDP, ATP, or ADP was added to reaction mixtures for polyPhe synthesis containing 0.5 m*M* GTP and assayed under conditions described (Bermek 1976). Final concentration of the nucleotides from the preincubation was 0.25 m*M*

Neverthless, the preincubation of EF-2 with oxidized GDP, and to a lesser extent with oxidized GTP, results in a certain decrease of EF-2 activity in polyPhe synthesis. The preincubation of EF-2 with GDP or ADP, but not with GTP or ATP, also results in a certain inhibition in polyPhe synthesis (Fig. 5).

An inhibition of polyPhe synthesis is, however, observed in parallel to an increase in the concentrations of GDP or ADP added directly to polyPhe synthesizing reaction mixtures (Fig. 6). GDP is most effective in this respect and causes at a concentration of ~0.4 m*M* GDP a 50% inhibition of protein synthesis. Pyrimidine nucleotide diphosphates or purine nucleotide triphosphates are here without any effect. The inhibition promoted by the increased concentrations of GDP seems to be reversable in the presence of saturating concentrations of EF-2 (Fig. 7).

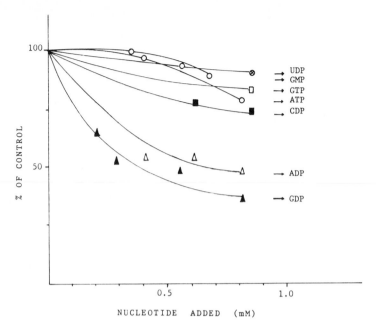

Fig. 6. Inhibiton of polyPhe synthesis by GDP and ADP. Increasing amounts of GTP, GDP, GMP, ATP, ADP, UDP, and CDP (0.2 mM to 1 mM) were directly added to reaction mixtures for polyPhe synthesis containing 0.5 mM GTP. The incubation was carried out for 10 min at 37°C under standard conditions (Bermek 1976). The results are expressed as the percentage of the control which contains only 0.5 mM GTP (=1.9 pmol Phe incorporated)

3 Discussion

The results reported imply that the binding site on EF-2 involved in direct interactions with guanine nucleotides is distinct from that involved in the ternary comlex formation. On account of our data we are inclined to assume that the binding site involved in the direct interactions with guanine nucleotides plays a modulating role on EF-2 activity. Alternatively, however, this site might be involved in a yet unknown function of EF-2, unrelated to protein synthesis. The former possibility is insofar attractive that it might provide a mechanism coupling the energy charge of the cell to polypeptide chain elongation rate. Previously, glucose deprival has been shown to reduce polypeptide chain initiation and elongation and the mechanism of this inhibition has remained till now obscure (van Venrooij et al. 1970). On account of the high affinity of GDP for eIF-2 and GDP-caused inhibition of the ternary initiation complex formation involving eIF-2. GTP. Met-tRNA^{fMet}, this complex has been regarded as the main step under the GDP-mediated control of cellular energy charge (Walton and Gill 1976a; Walton and Gill 1976b). However, the recently found eRF (Voorma 1984; Matts et al. 1984; Ochoa 1983) might minimize a regulatory effect through GDP on this step. The way of action of glucose deprival on chain elongation rate is equally unknown, but the possibility of the presence of a second nucleotide binding site on EF-2 will likely prompt further investigations which might provide insight into this control mechanism.

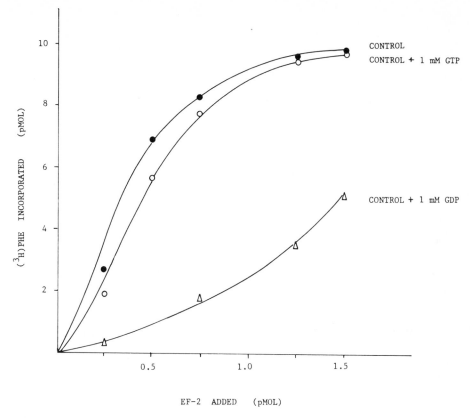

Fig. 7. Reversal of GDP caused inhibiton of polyPhe synthesis by EF-2. Increasing amounts of EF-2 (0 to 1.5 pmol) were added to reaction mixtures for polyPhe synthesis containing 0.5 mM GTP (control), 1.5 mM GTP (control + 1 mM GTP), and 0.5 mM GTP + 1 mM GDP (control + 1 mM GDP). PolyPhe synthesis was assayed under standard conditions (Bermek 1976) for 10 min at 37°C

Acknowledgement. This investigation was supported by a grant (TAG-479) from the Scientific and Technical Research Council of Turkey (TÜBITAK).

Abbreviations

EF-2, eukaryotic elongation factor 2; EF-G, prokaryotic elongation factor G; eRF, eukaryotic recycling factor; ADPR, adenosine diphosphate ribose; eIF-2, eukaryotic initiation factor 2; polyPhe, polyphenylalanine; GuoPP(CH$_2$)P, guanosine 5'-(β,γ-methylene) triphosphate; GDP$_{ox}$, GDP oxidized by NaIO$_4$ treatment; GDP$_{ox\text{-}red}$, GDP oxidized by NaIO$_4$ and subsequently reduced by incubation with NaBH$_4$; MalNEt, N-ethylmaleimide; TNBS, trinitrobenzenesulfonate; DMA, dimethyladepimidate; DMS, dimethylsuberimidate; SDS, sodium doedecyl sulfate.

References

Arai N, Arai K and Kaziro Y (1975) Formation of a binary complex between elongation factor G and guanine nucleotides. J Biochem (Tokyo) 78: 243-246

Bermek E and Matthaei H (1971) Interactions of human translocation factor with guanosine nucleotides and ribosomes. Biochemistry 10: 4906-4912

Bermek E (1976) Interactions of adenosine diphosphate-ribosylated elongation factor 2 with ribosomes. J Biol Chem 251: 6544-6549

Bermek E (1978) Mechanisms in polypeptide chain elongation on ribosomes. Prog Nucleic Acid Res Mol Biol 21: 63-100

Bielka H (1982) The eukaryotic ribosome. Springer, Berlin Heidelberg New York

Bodley JW, Lin L, Salas M and Tao M (1970) Studies on translocation. V. Fusidic acid stabilization of a eukaryotic ribosome-translocation factor-GDP complex, FEBS Lett 11: 153-156

Chuang DM and Weissbach H (1972) Studies on elongation factor II from calf brain. Arch Biochem Biophys 152: 114-124

Clark BFC (1980) The elongation step of protein synthesis. Trends Biochem Sci 5: 207-210

Collier RJ (1967) Effect of diphtheria toxin on protein synthesis: inactivation of one of the factors. J Mol Biol 25: 83-98

Girshovich AS, Pozdnyakov VA and Ovchinnikov YA (1976) Localization of the GTP-binding site in the ribosome-elongation factor G-GTP complex. Eur J Biochem 69: 321-328

Henriksen O, Robinson EA and Maxwell ES (1975a) Interaction of guanosine nucleotides with elongation factor 2. I. Equilibrium dialysis studies. J Biol Chem 250: 720-724

Henriksen O, Robinson EA and Maxwell ES (1975b) Interaction of guanosine nucleotides with elongation factor 2. II. Effect of ribosomes and magnesium ions on guanosine diphosphate and guanosine triphosphate binding to the enzyme. J Biol Chem 250: 725-730

Honjo T, Nishizuka Y, Hayaishi O and Kato J (1968) Diphtheria toxin dependent adenosine diphosphate ribosylation of aminoacyl transferase II and inhibition of protein synthesis. J Biol Chem 243: 3553-3555

Iglewski BH, Kabat D (1975) NAD-dependent inhibition of protein synthesis by Pseudomonas aeruginosa, toxin, Proc Natl Acad Sci USA 72: 2284-2288

Marsh RC, Chinali G and Parmeggiani A (1975) Function of sulfhydryl groups in ribosome-elongation factor G reactions. Assignment of guanine nucleotide binding site to elongation factor G. J Biol Chem 250: 8344-8352

Matts RL, Levin DH, Petryshyn R, Thomas NS and London IM (1984) this volume

Miller D, Weissbach H (1977) Aminoacyl-tRNA transfer factors. In: Weissbach H and Pestka S (eds) Molecular mechanisms of protein synthesis. Academic, New York pp 323-373

Mizumoto K, Iwasaki K and Kaziro Y (1974) Studies on polypeptide elongation factor 2 from pig liver. III.Interaction with guanine nucleotides in the presence and absence of ribosomes. J Biochem (Tokyo) 76: 1269-1280

Montanaro L, Sperti S and Mattioli A (1971) Interaction of ADP-ribosylated aminoacyl transferase II with GTP and ribosomes. Biochim Biophys Acta 238: 493-497

Nurten R and Bermek E (1980) Interactions of elongation factor 2 (EF-2) with guanine nucleotides and ribosomes. Binding of periodateoxidized nucleotides to EF-2. Eur J Biochem 103: 551-555

Nurten R, Bilgin-Aktar N and Bermek E (1983) Functional groups of elongation factor 2 involved in interactions with guanosine nucleotides and ribosomes. FEBS Lett 154: 391-394

Ochoa S (1983) Regulation of protein synthesis initiation in eukaryotes. Arch Biochem Biophys 223: 325-349

Raeburn S, Goor RS, Schneider JA and Maxwell ES (1968) Interaction of aminoacyl transferase II and guanosine triphosphate: inhibition by diphtheria toxin and nicotinamide adenine dinucleotide, Proc Natl Acad Sci USA 61: 1428-1434

Spirin AS and Ovchinnikov LP (1983) RNA-binding proteins involved in realization of genetic information in eukaryotic cells. Folia Biol (Praha) 29: 115-139

Van Ness BG, Barrowclough B and Bodley JW (1980a) Recognition of elongation factor 2 by diphtheria toxin is not solely defined by the presence of diphthamide. FEBS Lett 120: 4-6

Van Ness BG, Howard JB, Bodley JW (1980b) ADP-ribosylation of elongation factor 2 by diphtheria toxin. NMR spectra and proposed structures of ribosyl-diphthamide and its hydrolysis products. J Biol Chem 255: 10710-10717

Van Venrooij WJW, Henshaw EC and Hirsch CA (1970) Nutritional effects on the polyribosome distribution and rate of protein synthesis in ehrlich ascites tumor cells in culture, J Biol Chem 245: 5947-5953

Voorma HO (1984) this volume

Walton GM and Gill GN (1976a) Regulation of ternary (Met-tRNA$_f$. GTP. eukaryotic initiation factor 2) protein synthesis initiation complex formation by the adenylate energy change. Biochim Biophys Acta 418: 195-303

Walton GM and Gill GN (1976b) Preferential regulation of protein synthesis initiation complex formation by purine nucleotides, Biochim Biophys Acta 447: 11-19

Evolution and Selective Action of Translational Inhibitors

Some Interesting Features of a Halobacterial Gene and its mRNA

M. ŞIMŞEK[1]

1 Introduction

So far, little is known about the structure and transcription of genes in halobacteria. These bacteria are classified as "archaebacteria" by Woese and colleagues (Fox et al. 1980) and contain several features which are eukaryotic in nature. Thus, one wonders if the structural organization of halobacterial genes and mRNAs are different from the bacterial counterparts in some respects. For example, it would be interesting to see if introns are present in halobacterial genes. It would also be interesting to compare the structural features of promotors and the Shine-Dalgarno regions with the prokaryotic counterparts (Shine and Dalgarno 1974).

In this study, the gene for a membrane protein (bacteriorhodopsin) of *Halobacterium halobium* has been cloned and the nucleotide sequence of the structural gene with its flanking regions has been determined. Of the 1229 nucleotides sequenced, 786 correspond to the structural gene, 360 are upstream from the initiation codon and 83 are downstream from the COOH terminus. Comparison of the DNA sequence with the previously determined amino acid sequence of bacteriorhodopsin (Khorana et al. 1979) shows that there are no intervening sequences in the BR gene and that no prokaryotic type promotor or Shine-Dalgarno consensus sequence can be identified in the region upstream from the structural BR gene. Furthermore, the 5' terminus of the BR mRNA appears to contain only two or three nucleotides before the initiating AUG codon. Standard genetic code is used for the translation of the mRNA; however, there is a marked preference for either C or G in the third position of codons.

2 Synthesis and Cloning of a cDNA Fragment Which is Complementary to the NH$_2$-Terminal End of Bacteriorhodopsin Gene

A single-stranded cDNA which is complementary to the 5' terminus of the bacteriorhodopsin mRNA was synthesized using partially purified BR mRNA as template and the synthetic dodecanucleotide, d(C-C-A-G-A-T-C-C-A-C-T-C) as primer in a reverse trancriptase catalyzed reaction (Fig. 1). The cDNA synthesized was purified by electro-

---

[1]Middle East Technical University, Department of Biological Sciences, Ankara, Turkey

Mechanisms of Protein Synthesis
ed. by Bermek
© Springer-Verlag Berlin Heidelberg 1984

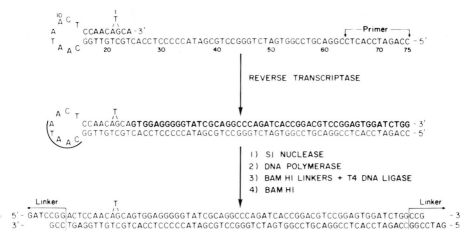

Fig. 1. Synthesis and cloning of double-stranded cDNA corresponding to the 5' terminus of bacteriorhodopsin gene. Single-stranded cDNA was synthesized using partially purified BR mRNA as template and a synthetic dodecanucleotide as primer: d(C-C-A-G-A-T-C-C-A-C-T-C). The primer sequence was derived from the amino acid sequence of bacteriorhodopsin at position 9 to 12 (.. Glu[9]--Trp[10]-11[11]-Trp[12]-). Synthesis of double-stranded cDNA was achieved by using avian myeloblastosis virus reverse transcriptase. The product was treated with S_1-nuclease and DNA polymerase I before the addition of BamH$_I$ linkers and ligation into the BamH$_I$ site of pBR$_{322}$

phoresis on an 8% polyacrylamide gel, and was used as template for the synthesis of double-stranded cDNA (ds-cDNA). Formation of a 9 basepairs(bp) hairpin loop structure at the 3' end of the c-DNA enabled the synthesis of ds-cDNA as shown in Fig. 1. After treatment of the ds-cDNA with S_1-nuclease, BamH$_I$ linkers were added and inserted into the BamH$_I$ site of pBR$_{322}$. Availability of such cloned cDNA served not only as a suitable probe for the cloning of the genomic BR gene, but it also provided the partial nucleotide sequence of the gene at the NH$_2$ terminal end.

3 Cloning and Characterization of a Pst$_I$ Fragment Which Contains the Intact Bacteriorhodopsin Gene

Total *H. halobium* DNA was digested with Pst$_I$ and inserted into the Pst$_I$ site of pBR$_{322}$. The recombinants obtained were screened by colony filter hybridization using the cloned cDNA insert (74 bp) described above as the probe. The gene was isolated on 5.2 kbp Pst$_I$ fragment. Digestion of this fragment with BamH$_I$ and BstE$_{II}$ yielded a 1.6 kbp fragment that retained the entire BR gene. The BamH$_I$-BstE$_{II}$ fragment was used for sequence analysis using a combination of the Maxam-Gilbert (1980) and the dideoxychain termination (Sanger et al. 1977) methods. Figure 2 shows the strategy for the sequence analysis and the position of some of the restriction sites on the 1.6 kbp fragment. The total nucleotide sequence of the gene is shown in Fig. 3. The sequence includes the structural gene (786 bp), together with 360 bp upstream from the

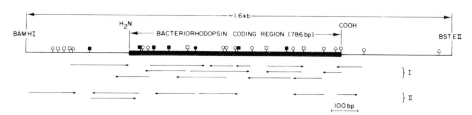

Fig. 2. Strategy for sequence analysis of the 1.6 kbp BamH$_I$-BstE$_{II}$ fragment. The *arrows* indicate the direction and extent of DNA sequences determined by (**I**) dideoxychazin termination and (**II**) Maxam–Gilbert methods. Restriction sites: ○, Sau3A; ■, Sau 96; ▢, TaqI; ●, KpnI; △, SmaI

initiating AUG codon and 80 bp downstream from the terminating codon. The derived DNA sequence is in complete agreement with the previously determined amino acid sequence of the mature bacteriorhodopsin. Furthermore, the gene sequence provides information for 13 extra amino acids at the NH$_2$ terminus which are not present in the mature bacteriorhodopsin. Thus, these additional amino acids must represent the precursor

```
GGGTGCAACCGTGAAGTCCGCCACGACCGCGTCACGACAGGAGCCGACCAGCGACACCCAGAAGGTGCGAACGGTTGAGTGCCGCAACGA
-360                                                                              -271

TCACGAGTTTTTCGTGCGCTTCGAGTGGTAACACGCGTGCACGCATCGACTTCACCGCGGGTGTTTCGACGCCAGCCGGCCGTTGAACCA
-270                                                                              -181

GCAGGCAGCGGGCATTTACAGCCGCTGTGGCCCAAATGGTGGGGTGCGCTATTTTGGTATGGTTTGGAATCCGCGTGTCGGCTCCGTGTC
-180                                                                               -91

TGACGGTTCATCGGTTCTAAATTCCGTCACGAGCGTACCATACTGATTGGGTCGTAGAGTTACACACATATCCTCGTTAGGTACTGTTGC
-90                                                                                 -1
```

```
     Met Leu Glu Leu Leu Pro Thr Ala Val Glu Gly Val Ser Gln Ala Gln Ile Thr Gly Arg Pro Glu Trp Ile Trp Leu Ala Leu Gly Thr
     ATGTTGGAGTTATTGCCAACAGCAGTGGAGGGGGTATCGCAGGCCCAGATCACCGGACGTCCGGAGTGGATCTGGCTAGCGCTCGGTACG
     1                                                                                               90

     Ala Leu Met Gly Leu Gly Thr Leu Tyr Phe Leu Val Lys Gly Met Gly Val Ser Asp Pro Asp Ala Lys Lys Phe Tyr Ala Ile Thr Thr
     GCGCTAATGGGACTCGGGACGCTCTATTTCCTCGTGAAAGGGATGGGCGTCTCGGACCCAGATGCAAAGAAATTCTACGCCATCACGACG
     91                                                                                             180

     Leu Val Pro Ala Ile Ala Phe Thr Met Tyr Leu Ser Met Leu Leu Gly Tyr Gly Leu Thr Met Val Pro Phe Gly Gly Glu Gln Asn Pro
     CTCGTCCCAGCCATCGCGTTCACGATGTACCTCTCGATGCTGCTGGGGTATGGCCTCACAATGGTACCGTTCGGTGGGGAGCAGAACCCC
     181                                                                                            270

     Ile Tyr Trp Ala Arg Tyr Ala Asp Trp Leu Phe Thr Thr Pro Leu Leu Leu Leu Asp Leu Ala Leu Leu Val Asp Ala Asp Gln Gly Thr
     ATCTACTGGGCGCGGTACGCTGACTGGCTGTTCACCACGCCGCTGTTGTTGTTAGACCTCGCGTTGCTCGTTGACGCGGATCAGGGAACG
     271                                                                                            360

     Ile Leu Ala Leu Val Gly Ala Asp Gly Ile Met Ile Gly Thr Gly Leu Val Gly Ala Leu Thr Lys Val Tyr Ser Tyr Arg Phe Val Trp
     ATCCTTGCGCTCGTCGGTGCCGACGGCATCATGATCGGGACCGGCCTGGTCGGCGCACTGACGAAGGTCTACTCGTACCGCTTCGTGTGG
     361                                                                                            450

     Trp Ala Ile Ser Thr Ala Ala Met Leu Tyr Ile Leu Tyr Val Leu Phe Phe Gly Phe Thr Ser Lys Ala Glu Ser Met Arg Pro Glu Val
     TGGGCGATCAGCACCGCAGCGATGCTGTACATCCTGTACGTGCTGTTCTTCGGGTTCACCTCGAAGGCCGAAAGCATGCGCCCCGAGGTC
     451                                                                                            540

     Ala Ser Thr Phe Lys Val Leu Arg Asn Val Thr Val Val Leu Trp Ser Ala Tyr Pro Val Val Trp Leu Ile Gly Ser Glu Gly Ala Gly
     GCATCCACGTTCAAAGTACTGCGTAACGTTACCGTTGTGTTGTGGTCCGCGTATCCCGTCGTGTGGCTGATCGGCAGCGAAGGTGCGGGA
     541                                                                                            630

     Ile Val Pro Leu Asn Ile Glu Thr Leu Leu Phe Met Val Leu Asp Val Ser Ala Lys Val Gly Phe Gly Leu Ile Leu Leu Arg Ser Arg
     ATCGTGCCGCTGAACATCGAGACGCTGCTGTTCATGGTGCTTGACGTGAGCGCGAAGGTCGGCTTCGGGCTCATCCTCCTGCGCAGTCGT
     631                                                                                            720

     Ala Ile Phe Gly Glu Ala Glu Ala Pro Glu Pro Ser Ala Gly Asp Gly Ala Ala Ala Thr Ser Asp  ***
     GCGATCTTCGGCGAAGCCGAAGCGCCGGAGCCGTCCGCCGGCGACGGCGCGGCCGCGACCAGCGACTGATCGCCACACGCAGGACAGCCCC
     721                                                                                             810
```

```
ACAACCGGCGCGGCTGTGTTCAACGACACACGATGAGTCCCCCACTCGGTCTTGTACTC
811                                                        869
```

Fig. 3. DNA sequence of the bacteriorhodopsin gene and its adjacent regions. *Arrows* indicate the NH$_2$ and COOH termini of the mature protein. Precursor region is composed of 13 amino acids which start with Met at position 1

region for bacteriorhodopsin. There is no other in phase ATG initiation codon before the precursor region, therefore, the size of the precursor appears to be only 13 amino acids long and this is quite shorter than the previously reported precursor sequences (Sabatini et al. 1982; Steiner et al. 1980).

The gene sequence also provides information for an additional amino acid (aspartic acid) at the COOH terminus which is not present in mature bacteriorhodopsin. The extra aspartic acid must have been removed during maturation of the protein. Presence of information for one additional amino acid has also been reported for α-tubulin (Valenzuela et al. 1981) and immunoglobulin heavy chain (Tucker et al. 1979) genes.

4 Interesting Features of the Bacteriorhodopsin Gene for Transcription

Very little is known about the transcription of genes in halobacteria. There are yet no concensus sequences available for the halobacterial promotors or transcriptional signals. Analysis of the BR gene sequence upstream from the ATG initiating codon shows that no prokaryotic type promotor sequence can be identified in this region (Fig. 3). In order to locate the position of the transcriptional start site for the bacteriorhodopsin gene, a nuclease-S_1 mapping experiment was performed using the hybridization between the BR mRNA and the separated strands of a 193 bp Sau 96 DNA fragment which covers the 5'-terminal region of the BR gene (nucleotides-148 to + 45). The result of this experiment is shown in Fig. 4 and indicates that there are only two or three nucleotides at the 5' end of the BR mRNA before the initiation AUG codon. Thus, the 5'-leader sequence of the mRNA is unusually short and the transcriptional start site is very close to the initiation codon. It is possible that the mRNA used in this study was not the original transcript and that it was produced by specific cleavage of a longer original transcript. However, this possibility has been ruled out recently by DasSarma et al. (1983) who showed that the original BR transcript contains only two nucleotides before the initiating AUG codon. Thus, if the BR promotor is upstream from the structural gene, it is expected to be not far away from the transcriptional start site. There-

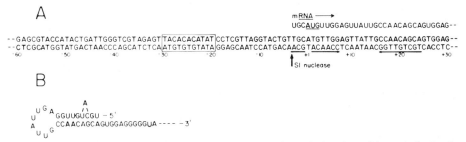

Fig. 4 A, B. A Sequence of DNA corresponding to the 5'-terminal region of bacteriorhodopsin mRNA. The initiating AUG codon is underlined in the mRNA. The *vertical arrow* shows the position beyond which DNA is protected from S_1-nuclease degradation by mRNA. The *boxed region* contains the alternating purine-pyrimidine stretch. The two *horizontal arrows* indicate the position of inverted repeat structure in DNA; **B** Harpin loop structure at the 5' end of bacteriorhodopsin mRNA

fore, the sequence of the BR gene has been compared (DasSarma et al. 1983) with the upstream sequences of the two halobacterial insertion elements (ISH_1, ISH_{50}) and the 5' end of the 5S RNA from *H. volcanii*. The only homology found is within the region 23 to 42 nucleotides upstream from the AUG start codon (Fig. 4). It is an A + T rich hexanucleotide (AAGTTA) of which at least five nucleotides are conserved in each case. It remains to be seen if this hexanucleotide plays a role as the transcriptional signal for halobacterial genes.

Another interesting feature noted in the 5'-terminal region of the BR gene is the presence of a 11 bp alternating purine-pyrimidine stretch which is located about 20 nucleotides upstream from the AUG start codon (Fig. 4). It is not known if this region plays any role for the transcription of BR gene. It is interesting to note that the salt concentration is very high (~4M KCl) in *H. halobium* and under such ionic conditions, DNA sequences that contain alternating purine-pyrimidine regions may adopt a Z-DNA structure (Wang et al 1979).

5 Interesting Features of the Bacteriorhodopsin mRNA

No mRNA has been isolated and fully characterized so far from halobacteria. Therefore, it is not known if halobacterial mRNAs contain features similar to those in other bacteria or if they differ in some respects.

Analysis of the BR-mRNA in this study by S_1-nuclease mapping showed that it contains only two or three nucleotides at the 5' end before the AUG codon (Fig. 4). In this respect it is strikingly different from prokaryotic mRNAs which usually contain 26 or more nucleotides at their 5'-leader regions (Steitz 1979). A purine rich sequence

UUU	Phe	0	UCU		0	UAU	Tyr	3	UGU	Cys	0
UUC		13	UCC	Ser	3	UAC		8	UGC		0
UUA	Leu	2	UCA		0	UAA	Ter	0	UGA	Ter	1
UUG		6	UCG		5	UAG		0	UGG	Trp	8
CUU		2	CCU		0	CAU	His	0	CGU		3
CUC	Leu	12	CCC	Pro	3	CAC		0	CGC	Arg	3
CUA		2	CCA		3	CAA	Gln	0	CGA		0
CUG		15	CCG		6	CAG		4	CGG		1
AUU		0	ACU		0	AAU	Asn	0	AGU	Ser	1
AUC	Ile	15	ACC	Thr	7	AAC		3	AGC		5
AUA		0	ACA		2	AAA	Lys	3	AGA	Arg	0
AUG	Met	10	ACG		10	AAG		4	AGG		0
GUU		3	GCU		1	GAU	Asp	2	GGU		4
GUC	Val	8	GCC	Ala	8	GAC		8	GGC	Gly	10
GUA		3	GCA		5	GAA	Glu	4	GGA		4
GUG		9	GCG		16	GAG		7	GGG		8

Fig. 5. Frequency of codon usage in the bacteriorhodopsin gene. Ter, termination

which is approximately 6 to 12 nucleotides upstream from the AUG codon is postulated by Shine-Dalgarno (1974) to interact with the complementary sequence at the 3' end of the prokaryotic 16S rRNA. Such an interaction is not possible for the BR mRNA, since it has an extremely short 5'-leading sequence. It is worth noting, however, that BR mRNA contains Shine-Dalgarno type sequence, GGAG (nucleotides 6-9), and GGAGG (nucleotides 27-31), but these are downstream from the initiating AUG codon.

Another interesting feature at the 5' end of the BR mRNA is the presence of a 9 bp interrupted inverted repeat which could fold to form a stable hairpin loop structure as shown in Fig. 4B. Such a structure has been found in several other mRNAs including the nifH mRNA of *Klebsiella pneumoniae*, ovalbumin mRNA (Mc Reynolds et al. 1978), and satellite tobacco necrosis virus RNA (Leung et al. 1979). Whether this structure plays a role in the functioning or stabilization of the mRNAs is not known.

A comparison between the structural gene and the amino acid sequence of the bacteriorhodopsin shows that standard genetic code is used by halobacteria (Fig. 5). The present work provides no information about assignments of codons for His and Cys which are absent in bacteriorhodopsin. There are also no AGA and AGG codons used for arginine. However, analysis of the published DNA sequence for the halobacterial insertion elements, ISH$_1$ (Şimşek et al. 1982), ISH$_2$ (DasSarma et al. 1983), and ISH$_{50}$ (Xu and Doolittle 1983) show that the normal codons are used for His, Cys, and Arg, in the long open reading frames of these elements.

There is a strong preference (82%) for the use of codons in bacteriorhodopsin ending in G or C. This is significantly higher than the overall G + C content (61.5%) of the coding region for the bacteriorhodopsin gene (Fig. 3). The G+C content of the total *H. halobium* DNA is reported to be approximately 67% by Moore and McCarthy (1969).

Abbrevations

kbp, kilo base pairs; ds-cDNA, double-stranded cDNA; BR, bacteriorhodopsin.

Acknowledgement. This work was carried out in Dr.H.G.Khorana's laboratory at MIT during the sabbatical leave of the author who deeply thanks Drs.H.G.Khorana and U.L.RajBhandary for their helpful discussions and S.DasSarma for making his manuscript available before publication.

References

DasSarma S, RajBhandary UL and Khorana HG (1983) Bacterio-opsin mRNA in wild type and bacterio-opsin deficient *Halobacterium halobium* strains. (submitted to Proc Natl Acad Sci USA)

DasSarma S, RajBhandary UL and Khorana HG (1983) High-frequency spontaneous mutation in the bacterioopsin gene in *Halobacterium halobium* is. mediated by transposable elements. Proc Natl Acad Sci USA 80: 2201-2205

Fox GE, Stackebrandt E, Hespel RB et al. (1980) The phylogeny of prokaryotes. Science (Wash DC) 209: 457-463

Khorana HG, Gerber GE, Herlihy WC, Gray CP, Anderegg RJ, Nihei K and Biemann K (1979) Amino acid sequence of bacteriorhodopsin. Proc Natl Acad Sci USA 76: 5046-5050

Leung DW, Browning KS, Heckman JE, RajBhandary UL and Clark JM (1979) Nucleotide sequence of 5' terminus of satellite tobacco necrosis virus ribonucleic acid. Biochemistry 18: 1361-1366

Maxam AM and Gilbert W (1980) Sequencing end-labelled DNA with base specific chemical cleavages. Methods Enzymol 65: 499-560

McReynolds L, O'Malley BW, Nisbet AD, Fothegill JE, Fields S, Robertson M and Brownlee GG (1978) Sequence of chicken ovalbumin mRNA. Nature (Lond) 273: 723-728

Moore RL and McCarthy BJ (1969) Characterization of the deoxyribonucleic acid of various strains of halophilic bacteria. J Bacteriol 99: 248-254

Sabatini D, Kreibich G, Morimoto T and Adesnick M (1982) Mechanisms for the incorporation of proteins in membranes and organelles. J Cell Biol 92: 1-22

Sanger F, Nicklen S and Coulson AR (1977) DNA sequencing with chain-termination inhibitors. Proc Natl Acad Sci USA 74: 5463-5467

Shine J and Dalgarno L (1974) The 3'-terminal sequence of E. coli 16S ribosomal RNA; complementarity to nonsense triplets and ribosome binding sites. Proc Natl Acad Sci USA 71: 1342-1346

Şimşek M, DasSarma S, RajBhandary UL and Khorana HG (1982) A transposable element from *Halobacterium halobium* which inactivates the Bacteriorhodopsin gene. Proc Natl Acad Sci USA 79: 7268-7272

Steiner DF, Quinn PS, Patzelt C, Chan SJ, Marsh J and Tager HS (1980) In: Goldstein L and Prescott DM (eds) Cell biology: a comprehensive treatise, vol. 4. Academic, New York, pp 175-201

Steitz JA (1979) In: Goldberg RF (ed) Biological regulation and development vol 1. Plenum, New York, pp 349-400

Tucker PW, Marcu KB, Slightom JL and Blattner FR (1979) Structure of the constant and 3' untranslated regions of the murine γ 2b heavy chain messenger RNA. Science (Wash DC) 206: 1299-1303

Valenzuela P, Quiroga M, Zaldivar J, Rutter JW, Kirschner MW and Cleveland DW (1981) Nucleotide and corresponding amino acid sequences encoded by α and β tubulin mRNAs, Nature (Lond) 289: 650-655

Wang AH, Quigley GJ, Kolpak FJ, Crawford JL, vanBoom JH, vander Marel G and Rich A (1979) Molecular structure of a left-handed double helical DNA fragment at atomic resolution. Nature (Lond) 282: 680-686

Xu WL and DooLittle UF (1983) Structure of the archaebacterial transposable element ISH$_{50}$. Nucleic Acids Res 11: 4195-4199

Cell Evolution and the Selective Action of Translation Inhibitors

D. VÁZQUEZ[1]

1 Introduction

The classification of all types of living cells into two ample groups, including the eukaryotic and the prokaryotic cells, was universally accepted a few years ago. Three fundamental postulates, as the bases for cellular evolution, were also widely accepted. The first one was that after a very long prebiotic period preceding the existance of real life, the first living cells were essentially of the type of the present prokaryotes (bacteria and blue-green algae). The second of these fundamental notions was that cellular evolution was essentially monophyletic. The third notion was really a consequence of the other two and recognized that all the present living cells were descendants from the primitive prokaryotes. There was the problem in these postulates of explaining the appearance of the eukaryotes descending from the prokaryotes in a relatively recent period. Furthermore, ample experimental evidence obtained in the sixties allowed to propose the endosymbiotic origin of mitochondria and chloroplasts in the eukaryotic cells. Further advances in the seventies confirmed the proposal widely accepted now that the genome of mitochondria and chloroplasts are descendants of the genome of aerobic bacteria and blue-green algae, respectively, living in symbiosis in the eukaryotic cell.

The above three postulates, considered as fundamental 10 years ago, are now very difficult to maintain. This is partly due to recent studies with microfossils which suggest that eukaryotic cells are really much oder than expected (Knoll and Barghoorn 1977). On the other hand, on the basis of recent studies on the chemical structure of different cellular components and on the mode and selective action of antibiotics, it was concluded that the new "Kingdom of Archaebacteriae" should be included within the prokaryotes since it is quite distinct from either eukaryotes or the other known prokaryotes (bacteria and blue-green algae) (Woese and Fox 1977; Fox et al. 1980; Doolittle 1980; Woese 1981; Matheson and Yaguchi 1982; Kandler 1982). Hence, the existence, within the prokaryotes, of two phylogenetically independent kingdoms, including the true bacteria or "eubacteriae" (bacteria and blue-green algae) in one of them, and the "archaebacteriae" in the other, was proposed. On the other hand, it was proposed that the present eukaryotic cells might have arisen from the symbiosis of an eubacteria (bacteria or blue-green algae originating mitochondria or chloroplasts, respectively) and a cell of a third hypothetic kingdom the "urkaryote" which preceded the nucleocytoplasmic structures of the present eukaryotic cell (Woese and Fox 1977; Fox et al. 1980; Woese

---

[1] Centro de Biologia Molecular CSIC and UAM Canto Blanco, Madrid 34, Spain

Mechanisms of Protein Synthesis
ed. by Bermek
© Springer-Verlag Berlin Heidelberg 1984

1981). Alternatively, these nucleocytoplasmic structures of the eukaryotic cell might have their origin in the archaebacteriae without the existence of the ancient kingdom of the urkaryotes which has already disappeared (Woese and Fox 1977; Fox et al. 1980).

There are some semantic objections to the above proposals as clearly pointed out (Doolittle 1980). Thus, the new term of eubacteria which includes bacteria and blue-green algae is conflicting with the present taxonomic term of "Eubacteria" which includes a more limited and well-defined group of bacteria. On the other hand, although the term archaebacteriae is widely accepted, it sounds like an unrealistic term suggesting that those microorganisms are living fossils; to avoid this semantic anomaly, the alternative term of metabacteriae (Osawa and Hori 1979; Hori et al. 1982) has been proposed for the archaebacteriae, but is not much used.

Another remarkable feature of the archaebacteriae is their enormous heterogeneity, biochemically and ecologically. Thus, three groups are distinguished at least in the archaebacterial kingdom, including (a) the halophylic (aerobic and requiring high salt concentrations), (b) methanogenic (strictly anaerobic synthesizing microorganisms), and (c) thermoacidophylic (growing at $75°$-$90°C$ in very acidic media under pH 2.0). There might be many more archaebacteria not yet isolated since the number of genera and species recently described is continuously increasing.

Moreover, at least two morphologically distinct microorganisms have been isolated recently growing at $350°C$ in waters emanating from sulfide chimneys in the Pacific Ocean and grown at the laboratory at $250°C$ (Baross and Deming 1983). We do not yet know if these microorganisms might be included within the archaebacteria or, having a number of new macromolecules with biological activities at such high temperatures, they will have to be classified in a new still undescribed kingdom.

2 Results

2.1 Protein Synthesis in Bacterial-type Systems. Site and Mode of Action of Protein Synthesis Inhibitors

A general scheme summarizing the mechanism of translation and site of action of inhibitors in bacterial systems is shown in Fig. 1. The scheme is essentially valid for bacteria (Gram positive, Gram negative, and acid-alcohol resistant), blue-green algae, chloroplasts, and mitochondria. However, it is known that there are some small differences in antibiotic sensitivity in some of the systems of the prokaryotic type. Thus, fusidic acids is inactive on ribosomes from *Neurospora crassa* mitochondria (Grandi et al. 1971) and in protein synthesizing systems from sporulating *Bacillus subtilis* (Fortnagel and Freese 1977). Furthermore, mitochondrial ribosomes appear to be less sensitive than bacterial ribosomes to a number of aminoglycoside and macrolide antibiotics (Vázquez 1979; review). This is not surprising since the long period of symbiosis of mitochondria in eukaryotic cells, unlikely in the shorter period of association of chloroplasts in the eukaryotes, has resulted in the shortening of the largest rRNA of both ribosomal subunits and the absence of the 5SRNA in their ribosomes.

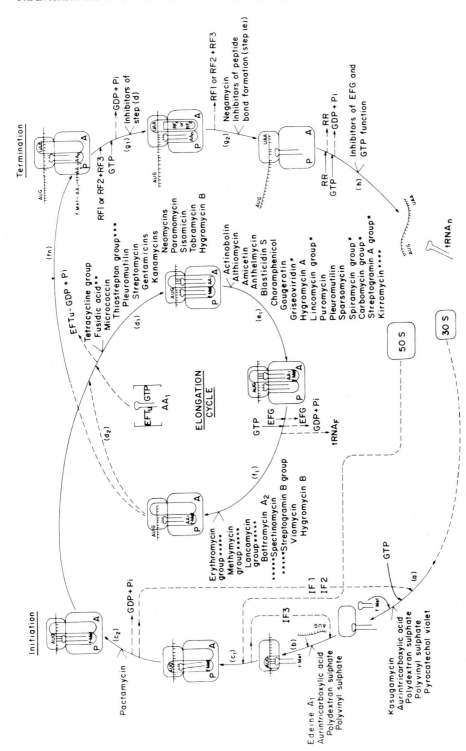

Fig. 1. Translation process in bacteria. Site of action of translation inhibitors. (Taken from Vázquez 1979)

Table 1. Antibiotic activity on Gram positive and Gram negative bacteria and blue-green algae[a]

Antibiotic	Antibiotic concentration (μM) inhibiting by 50% peptide bond formation			
	Escherichia coli	*Bacillus subtilis*	*Anacystis montana*	*Saccharomyces cerevisiae*
Chloramphenicol	10	10	10	>500
Streptogramin A	3	1	2	>500
Lincomycin	6	0.2	2	>500
Sparsomycin	20	20	20	20

[a]Peptide bond formation was studied in cell-free systems following Ac-Leu-puromycin formation by ribosomes using CACCA-Leu-Ac and ^{3}H-puromycin as substrates (Celma 1971; Battaner and Vázquez 1971)

The well-known differences in antibiotic sensitivity of some Gram positive and Gram negative bacteria are based in some cases not only in the permeability barrier, but also in a differential sensitivity of the ribosomes (Table 1) (Celma 1971, Battaner and Vázquez 1971). Thus, ribosomes of yeast, *B. subtilis, E. coli*, and the blue-green algae *A. montana* are equally sensitive to sparsomycin essentially in agreement with results obtained in intact microorganisms since the antibiotic has a very broad spectrum acting on prokaryotic and eukaryotic cells. Furthermore, *B. subtilis, E. coli*, and *A. montana* ribosomes are equally sensitive to chloramphenicol since the antibiotic has a broad spectrum, but acts only on prokaryotic ribosomes. On the other hand, streptogramin A and lincomycin are preferentially active on ribosomes from Gram positive bacteria and to a lesser extent on blue-green algae as clearly shown in Table 1, where the results obtained with *B. subtilis, E. coli,* and *A. montana* are compared. Indeed, binding experiments have shown the very low affinity of lincomycin for *E. coli* ribosomes in the absence of ethanol (Fernández-Munoz et al. 1971). However, the above examples illustrate the few exceptions in antibiotic sensitivity in different prokaryotic systems and show that evolution within bacterial and their derived systems for translation is fairly conservative.

2.2 Protein Synthesis in Eukaryotic Systems. Site and Mode of Action of Protein Synthesis Inhibitors

A general scheme, summarizing the mechanism of translation and site of action of inhibitors in eukaryotic systems is shown in Fig. 2. The scheme is essentially valid for all systems of translation in the cytoplasm of eukaryotic cells and there are really very few exceptions. Perhaps the most interesting one is the observation that tenuazonic acid does not act on yeast ribosomes and is mainly active on mammalian ribosomes (Carrasco and Vázquez 1973). The preservation of the eukaryotic systems for translation in evolution as judged by the selective action of inhibitors is indeed really astonishing. Furthermore, it can be seen by comparing Fig. 1 and 2 that there are a number of inhibitors which are equally active on bacterial and eukaryotic systems of translation. This is not surprising since both systems have a number of features in common although they differ in others.

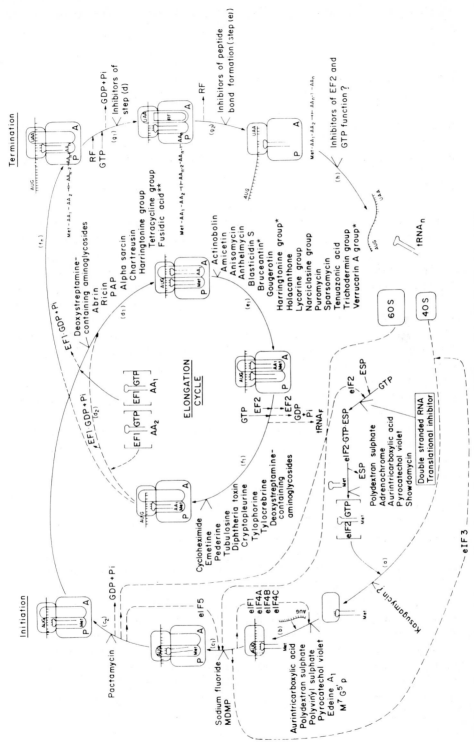

Fig. 2. Translation process in eukaryotic cells. Site of action of translation inhibitors (Taken from Vázquez 1979)

2.3 Protein Synthesis in Archaebacteria. Inhibitors of Translational in Archaebacterial Systems

The mechanism of translation in archaebacterial systems has not been elucidated. However, there are a number of features suggesting that the overall mechanism of translation in archaebacteria is similar to those of bacteria and eukaryotes since cell-free systems for polyphenylalanine synthesis and peptide-bond formation have been developed in *Halobacterium* (a halophylic archaebacteria) (Kessel and Klink 1981) and *Thermoyplasma* (a thermoacidophylic archaebacteria) (Kwok and Wong 1980). On the other hand, it is known that in archaebacteria the initiator substrate is Met-tRNA (Woese 1981; Kwok and Wong 1980) and the elongation factor involved in translocation is related to the EF-2 of eukaryotes having the characteristic basic amino acid diphthamide, unlike the bacterial corresponding EF-G factor (Kessel and Klink 1981; Kessel and Klink 1980; Pappenheimer et al. 1983). Although data concerning the mechanism of translation in archaebacteria are rather incomplete and fragmented there is much more information concerning the archaebacterial ribosome. (Kandler 1982; Osawa and Hori 1979; Hori and Osawa 1982; Matheson et al. 1980; Matheson and Yaguchi 1982). These data show that although archaebacterial ribosomes, ribosomal subunits, and ribosomal RNAs are similar to the bacterial ones in their sedimentation coefficients, they are more similar to the eukaryotic ones in a number of features, including the sequences of their A protein and ribosomal RNAs (Osawa and Hori 1979; Hori and Osawa 1982; Matheson et al. 1980; Matheson and Yaguchi 1982). On the other hand, some features of the archaebacterial ribosomes and their components, including the acidic nature of most of their ribosomal proteins, suggest that they are quite distinct from the eukaryotic and the bacterial ones (Matheson et al. 1980; Matheson and Yaguchi 1982).

We do not have enough data concerning the selective action of the antibiotics on archaebacterial ribosomes. Moreover, inhibitors selectively blocking translation by archaebacteria have not been described. However, we have a number of data concerning inhibitors blocking translation by archaebacteria and other types of systems (Table 2). Unfortunately, data presented in Table 2 have been obtained by different groups working with different systems and under different experimental conditions. Thus, the inhibitory effect of α-sarcin has been observed only in *Sulfolobus* (a thermoacidophilic archaebacteria) (Amils 1983; personal communication) the inhibitory effect of thiostrepton and streptogramin (synonims virginiamycin), shown in methanogenic cell-free systems (Elhart and Bock 1983) has not been confirmed with *Sulfolobus* ribosomes (Acca et al. 1983) and the reported inhibitory effect of edeine in *Sulfolobus* (Acca et al. 1983) should be confirmed in other systems. However, it is interesting that the data at our disposal suggest that ribosomal diversity within archaebacterial systems for translation is going to be higher than the eukaryotic and even the bacterial ones including their organellae-derived systems. Altogether data presented in Table 2 suggest that the overall mechanism of peptide-bond formation in archaebacteria is essentially similar to other known systems since puromycin is also an acceptor substrate. The peptidyl transferase center has some site(s) rather common to bacteria (sensitivity to streptogramin) or to eukaryotes (sensitivity to anisomycin). The step of aminoacyl-tRNA recognition might have in archaebacterial ribosomes some features similar to

Table 2. Selective action of translation inhibitors on cell-free systems[a]

Inhibitors acting on bacterial and archaebacterial systems	Inhibitors acting on eukaryotic and archaebacterial systems	Inhibitors acting on bacterial, archaebacterial, and eukaryotic systems
Thiostrepton (Elhart and Bock 1983) Streptogramin (Elhart and Bock 1983; Hilpert et al. 1981)	Anisomycin (Kessel and Klink 1981; Elhart and Bock 1983; Pecher and Bock 1981; Searcy 1982; Schmid et al. 1982) α-sarcin (Amils 1983; personal communication) Diphtheria toxin (Kessel and Klink 1981; Kessel and Klink 1980; Pappenheimer et al. 1983)	Puromycin (Kessel and Klink 1981) (Edeine) (Acca et al. 1983)

[a] References within brackets refer to those in which the pertinent data presented in this Table were reported. The data concerning triostrepton and streptogramin were obtained with cell-free systems from methanogenic archaebacteria, but were not confirmed in *Sulfolobus* (Acca et al. 1983). Chloramphenicol was reported to be active in intact cells of some species of *Halobacterium* and methanogenic archaebacteria (Hilpert et al. 1981; Pecher and Bock 1981), but not confirmed in cell-free systems (Elhart and Bock 1983; Acca et al. 1983; Schmid et al. 1982); it was postulated that the antibiotic inhibits in intact methanogenic archaebacteria the hydrogenases involved in methane production (Hilpert et al. 1981). Lack of sensitivity of archaebacterial ribosomes to a number of antibiotics acting on bacterial and eukaryotic systems has been reported (Elhart and Bock 1983; Acca et al. 1983; Cammarano et al. 1982)

bacteria, as judged by their sensitivity to thiostrepton or to eukaryotes, as judged by their sensitivity to α-sarcin. Translocation and the elongation factor involved in translocation in archaebacterial systems are closely related to the eukaryotic systems as clearly shown by their sensitivity to diphtheria toxin (Kessel and Klink 1981; Kessel and Klink 1980; Pappenheimer et al. 1983). We do not yet know how the reported sensitivity of *Sulfolobus* to edeine is relevant (Acca et al. 1983). However, since edeine is an inhibitor of initiation in bacterial and eukaryotic ribosomes (Vázquez 1979) and Met-tRNA is the initiator substrate, it appears that essentially the initiation process is not very different from those reactions already studied in bacteria and eukaryotes (Figs. 1 and 2).

Particularly interesting is the action of α-sarcin on archaebacterial ribosomes. This toxin is catalytically active mainly on eukaryotic ribosomes since it cleaves the 3'-terminal end of the larger ribosomal RNA of the larger ribosomal subunit (Schindler and Davies 1977; Conde et al. 1978) in a very well-defined site of the sequence which distance to the 3'-terminal end depends on the ribosomal-type (Endo and Wool 1982). Variation in a simple base on the RNA sequence of bacterial ribosomal RNA results in a sensitivity of bacterial ribosomes two orders of magnitude lower than the eukaryotic ribosomes (Endo and Wool 1982). Two other toxins, mitogillin and restrictocin, act on yeast and rabbit reticulocyte ribosomes similarly to α-sarcin (Conde et al. 1978). α-Sarcin. mitogillin, and restrictocin are very similar as shown by their mode of action and immunological cross-reactivity (Conde et al. 1978). However, α-sarcin differs from mitogillin and restrictocin as shown by immunological experiments in diffusion plates

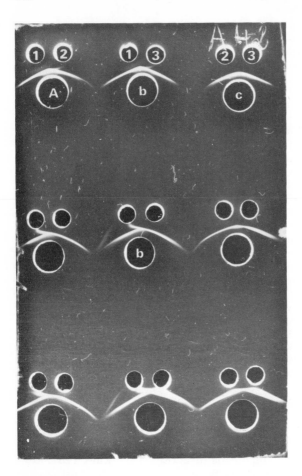

Fig. 3. Cross-reaction of α-sarcin, restrictocin, and mitogillin antisera with the three toxins by double diffusion tests. Wells *A, b,* and *c* contain 15 μl of α-sarcin, restrictocin, and mitogillin antisera, respectively. Wells *1, 2,* and *3* contain 3.0 μg of α-sarcin, restrictocin, and mitogillin, respectively (Conde, Vidal and Vázques, unpublished results)

using antisera for the three toxins (Fig. 3), whereas no differences can be observed by using these methods between mitogillin and restrictocin. Indeed α-sarcin structure has been elucidated (Sacco et al. 1983). It has been further shown that mitogillin and restrictocin differ between themselves in a single amino acid of their sequence, but both of them differ with α-sarcin in 15% of their amino acid sequence (Gavilanes et al. 1983; Méndez 1983, personal communication). However, α-sarcin is very active in *Sulfolobus* ribosomes, but no effects were observed by similar concentrations of mitogillin and restrictocin in this type of ribosome (Amils 1983, personal communication). This differential selectivity of the three toxins is very interesting indeed, although mitogillin and restrictocin might be active on archaebacterial ribosomes at higher concentrations, since these toxins can also cleave the 3'-terminal end of the larger RNA of the larger ribosomal subunit under certain experimental conditions (Amils 1983, personal communication) similarly as observed with α-sarcin on bacterial ribosomes.

3 Discussion

It is very interesting that despite the apparently drastic differential structure of archae-bacterial ribosomes, most of their proteins having an acidic nature, the overall mechanism of protein synthesis is not very different from that of bacterial and eu-karyotic systems. With respect to antibiotic inhibitors of translation, archaebacteria have a number of them common with bacteria and others similar to those of eukaryotes. Indeed, the data presently available and summarized above, concerning the selective action of inhibitors on archaebacterial ribosomes, support other experimental evidence favoring the proposal of archaebacteria as a third and independent kingdom quite dis-tict from both eukaryotes and bacteria.

References

Acca M, Londei P, Teicher A, del Piano L, Nicolaus B (1983) A thermoacidophilic archaebacteria intensitive to most prokaryotic and eukaryotic targeted antibiotics. 15th FEBS Meeting, Brussels 29-39 July. Abstracts S-00, WE-245, p 318

Baross JA, Deming JW (1983) Growth of "black smoker" bacteria at temperatures of at least 250°C. Nature (Lond) 303: 423-426

Battaner E, Vázquez D (1971) Inhibitors of protein synthesis by ribosomes of the 80S-type. Biochim Biophys Acta 154: 316-330

Cammarano P, Teichner A, Chinali G, Londei P, de Rosa M, Cambacorta A, Nicolaus B (1982) Ar-chaebacterial elongation factor Tu insensitive to pulvomycin and kirromycin. FEBS Lett 148: 255-259

Carrasco L, Vázquez D (1973) Differences in eukaryotic ribosomes detected by the selective action of an antibiotic. Biochim Biophys Acta 319: 209-215

Celma ML (1971) Ph D Thesis. Faculty of Pharmacy, Madrid University, Madrid, Spain

Conde FP, Fernandez-Puentes C, Montero MTV, Vázquez D (1978) Studies on the mode of action of alpha sarcin antibodies. FEMS Microbiol Lett 4: 349-355

Doolittle WF (1980) Revolutionary concepts in evolutionary cell biology. Trend Biochem Sci 5: 146-149

Elhart D, Bock A (1983) An in vitro polypeptide synthesizing system from methanogenic bacteria-sensitivity to antibiotics. Mol Gen Genet 188: 128-134

Endo Y, Wool I (1982) The site of action α-sarcin on eukaryotic ribosomes. The sequence at the α-sarcin cleavage site in 28S ribosomal ribonucleic acid. J Biol Chem 257: 9054-9060

Fernandez-Muñoz R, Monro RE, Torres-Pinedo R, Vázquez D (1971) Substrate -and antibiotic-binding sites at the peptidyl-transferase centre of E. coli ribosomes. Studies on the chloram-phenicol, lincomycin and erythromycin sites. Eur J Biochem 23: 185-193

Fortnagel P, Freese EB (1977) Morphological stages of Bacillus-subtilis sporulation and resistance to fusidic acid. J Gen Microbiol 101: 299-306

Fox GE, Stackebrandt E, Hespell RB et al. (1980) The phylogeny of prokaryotes. Science (Wash DC) 209: 457-463

Gavilanes JG, Vázquez D, Soriano F, Mendez E (1983) Chemical and spectroscopic evidence on the homology of three antitumor proteins: α-sarcin, mitogillin and restrictocin. J Prot Chem 2: 251-261

Grandi M, Hels A, Kuntzel H (1971) Fusidic acid resistance of mitochondrial G-factor from neuro-spora crassa. Biochem Biophys Res Commun 44: 864-871

Hilpert R, Winter J, Hammes W, Kandler O (1981) The sensitivity of archaebacteria to antibiotics. Zentralbl Bakteriol Abt I Orig C2 Hyg Krankenhaushyg Betriebshyg Praer Med 11-20

Hori H, Itoh T, Osawa S (1982) The phylogenetic structure of metabacteria. Zentralbl Bakteriol
    Abt I Orig C3 Hyg Krankenhaushyg Betriebshyg Praer Med 18-30
Kandler O (Ed) (1982) "Archaebacteria" (Chapters by different authors). Fischer, Stuttgart
Kessel M, Klink, F (1980) Archaebacterial elongation factor ADP-ribosylated by diphtheria toxin.
    Nature (London) 287: 250-251
Kessel M, Klink F (1981) Two elongation factors from the extremely halophilic archaebacterium
    halobacterium cutirubrum. Assay systems and purification at high salt concentrations. Eur J
    Biochem 114: 481-486
Knoll AH, Barghoorn ES (1977) Archean microfossils showing cell division from swaziland system
    of South Africa. Science (Wash DC) 198:396-398
Kwok Y, Wong JT (1980) Evolutionary relationship between halobacterium cutirubrum and eu-
    karyotes determined by use of aminoacyl-tRNA synthetases as phylogenetic probes. Can J
    Biochem 58: 213-218
Matheson AT, Moller W, Amons R, Yaguchi M (1980) In: Chambliss G, Craven GR, Davies J, Davis
    K, Kahan L, Nomura M (eds) Ribosomes. University Park Press, Baltimore, pp 297-332
Matheson AT, Yaguchi M (1982) The evolution of the archaebacterial ribosome. Zentralbl Bakeriol
    Abt I Orig C3 Hyg Krankenhaushyg Betriebshyg Praer Med 192-199
Osawa S, Hori H (1979) In: Chambliss G et al. (eds) Ribosomes. University Park Press, Baltimore,
    pp 335-355
Pappenheimer AM, Dunlop PC, Adolph KW, Bodley JW (to be published 1983) J Bacteriol
Pecher T, Bock A (1981) In vivo susceptibility of holophilic and methanogenic organisms to protein
    synthesis inhibitors. FEMS Microbiol Lett 10: 295-297
Sacco D, Drickamer K, Wool I (to be published 1983) J Biol Chem
Searcy DG (1982) Thermoplasma a primordial cell from a refuse pile. Trends Biochem Sci 7: 183-
    185
Schindler DG, Davies JE (1977) Specific cleavage of ribosomal RNA caused by alpha sarcin. Nucl
    Acids Res 4: 1097-1110
Schmid G, Pecher Th, Bock A (1982) Properties of the translational apparatus of archaebacteria.
    Zentralbl Bakteriol Abt I Orig C3 Hyg Krankenhaushyg Betriebshyg Praer Med 209-217
Vázquez D (1979) Inhibitors of protein biosynthesis. Springer, Berlin Heidelberg New York
Woese CR, Fox GE (1977) Phylogenetic structure of the prokaryotic domain: the primary king-
    doms. Proc Natl Acad Sci USA 74: 5088-5090
Woese CR (1981) Archaebacteria. Sci Am 244: 94-107

Concluding Remarks

The first session of the meeting focused on ribosome structure. The data presented in this session emphasized once again that contrary to the previously prevailing view the ribosomal RNA may fulfill active roles in different ribosomal functions and is more than a framework to which ribosomal proteins are attached. Thus, information obtained from structural work on the ribosomal RNA is apt to provide further, if not ultimate, insight into the ribosomal mechanisms of protein synthesis. Cantor summarized the results obtained by the electron microscopic analysis of long distance psoralen cross-links on *E. coli* 16S RNA with established orientation. From the estimated distances of the beginning and end of the cross-linked loop from the 5' end plotted as points in a triangular field, a pattern starts to emerge now, in general confirmatory, of the secondary structure models proposed for the 16S RNA. Additional cross-links observed by this approach, but thus far not considered in the secondary structure models are suggestive of the contacts determining the tertiary structure of the RNA molecule. The diversity of the contacts made by the 3' end of 16S RNA deserves here particular attention since it may attest to structural rearrangements of the 3' end in different steps of protein synthesis. The identification of the sequences involved in the cross-links is desirable for the test of secondary structure models and the construction of three-dimensional models. Cantor described a strategy to determine such sequences based on the hybridization of the RNA samples enriched by electrophoresis in particular cross-links to appropriate endlabeled complementary DNA fragments and subsequent exposure of the RNA-DNA hybrid to single-strand specific mung bean nuclease digestion.

An example of a possible involvement of the ribosomal RNA in a specific ribosomal function has been indeed provided by Küchler and co-workers. These investigators have shown the localization of a photoaffinity label specific for peptidyl transferase on the 23S RNA and identified the photoaffinity labeled residue(s) on the RNA molecule as U2584 (or U2585) in the central loop of domain V of the RNA molecule. This domain, corresponding to a sequence highly conserved throughout evolution, is known for being the site of mutations conferring resistance to erythromycin and chloramphenicol. The possibility of the 23S as a constituent of peptidyl transferase offers an alternative to the views tending to explain this activity in terms of ribosomal proteins with catalytic properties (or contributing together to the formation of such catalytic activity).

Another approach for elucidating the ribosomal structure and the structure-function relations in protein synthesis is based on fluorescence techniques. Hardesty gave a survey of the distances between different ribosomal components and between the ribosomal components and the fluorescent labeled tRNA molecule or EF-Tu measured by energy transfer experiments. His data show the 3' end of the 16S RNA molecule to be, consistent with previous evidence, located closer to the tRNA bound to P-site than that bound to A-site, and moreover, closer to the central region of the P-site bound tRNA than its 3' end or decoding loop. The energy transfer experiments with the coumarin derivative of EF-Tu suggest that it may bind on the ribosome to a region at the base of the L7/ L12 stalk. Moreover, a considerable distance is shown to exist between fluorescent labeled S1 between its site involved in the binding of mRNA) and A-site bound Phe-tRNA^{Phe} labeled at its Y-base with proflavine. In other words, a considerable length of mRNA spans the gap between the polynucleotide binding site of S1 and the decoding site of the 30S subunit.

Additional data on the relative arrangement of A- and P-site bound tRNA^{Phe} molecules has been obtained by Wintermeyer and co-workers also employing energy transfer measurements with different donor-acceptor pairs. The results reveal close arrangement of the anticodons and moderate distance between D loops of the two RNA molecules. Provided that both tRNA molecules preserve the tertiary structure present in crystal and that the 3' ends are as close as anticipated, they conclude the data would be consistent with the models where the two tRNAs with a somewhat asymmetric arrangement of their planes form an angle of $60° \mp 30°$.

In the elongation session Parmeggiani reported findings revealing some new aspects of the mechanism of action of EF-Tu, which has attracted in recent years considerable attention due to its multifunctional properties. Parmeggiani has studied the interaction between EF-Tu and aa-tRNA (or its fragments) using GTP hydrolysis in the presence of limiting amounts of EF-Tu as a probe. In the presence of kirromycin, aa-tRNA, and to lesser degrees its 3'-end fragments show stimulatory effects on EF-Tu catalyzed GTP hydrolysis with or without ribosomes. Kirromycin is known to activate the EF-Tu site for GTP hydrolysis in the absence of ribosomes or aa-tRNA by relieving the constraints keeping this activity in its physiological limits. Remarkably, in the absence of kirromycin aa-tRNA (and the fragment corresponding to its 3'-half) display an inhibitory effect on EF-Tu catalyzed GTP hydrolysis on ribosomes (or without ribosomes) with increasing NH_4Cl and decreasing $MgCl_2$ concentrations. The inhibition is most marked at $MgCl_2$ concentrations below 10 mM. In contrast, aminoacylated 3'-end fragments, i.e., CCA-aa or CA-aa, stimulate the reaction with increasing NH_4Cl concentrations and this stimulatory effect is more or less unaffected by $MgCl_2$ concentrations used. Different effects of the whole (or 3'-half) aa-tRNA molecule vs its short 3' fragments suggest that besides the fundamental role of the aminoacylated 3' end in the activation of the catalytic center for GTP hydrolysis, certain regions (possibly TψC loop and stem and the adjacent regions) exert an additional control effect on this activity. On the other hand, truncated tRNA in the presence of CCA-aa fragments does not reveal the inhibitory effects which aa-tRNA by itself exerts. Thus, EF-Tu may induce in its turn conformational changes at aa-tRNA influencing the orientation of the 3' extremity.

Sprinzl has employed, on the other hand, specific chemical modifications of aa-tRNAs and guanosine nucleotides, introduction of spectroscopic labels into tRNA molecules followed by physicochemical measurements and affinity chromatography on immobilized EF-Tu-GTP to study the interactions of EF-Tu. He reported that EF-Tu requires for the optimal recognition of amonoacyl-tRNA the presence of a protonated aminoacyl residue which is likely attached in an ortoacid form to both 2'- and 3'-hydroxyl groups of the terminal adenosine. Since base substitutions on positions 74 and 75 do not impair the ternary complex formation, argued Sprinzl, the nucleobases of the CCA-aa end may not interact directly with EF-Tu. Thus, the nucleobases of the 3' terminus of aminoacyl-tRNA may be involved in an interaction with the peptidyl transferase A-site after the ternary complex binds to the ribosome (as has been also discussed by Küchler). Measurements of ESR spectra of the EF-Tu complexes involving guanine nucleotides spin-labeled at 3' amino group reveal aa-tRNA dependent changes in the mode of attachment of the nucleotide to EF-Tu. The unimpaired activity of spin-labeled nucleotides seem to indicate that the ribose moiety of the bound nucleotides is orientated towards the solvent.

The conventional model assumes the presence of two tRNA binding sites on the ribosome. However, repeatedly, alternatives to this model with three or more binding sites have been proposed. According to some of these models an additional binding site precedes the A-site and has been denoted as entry or R (recognition) site. Still others have reported the existence of a third [exit(E) or discharge (D)] site which, accessible only for deacylated tRNA, receives the deacylated tRNA from the P-site during translocation and represents at the same time the site of its release from the ribosome. Wintermeyer reported in the meeting the kinetic data obtained by fluorescence stopped-flow experiments supporting the existence of the D-site. However, since the dissociation of tRNA from the D-site is fast at Mg^{2+} concentrations of 10 mM and below, the physiological significance of this third site remains obscure. Kinetic data show further that binding of deacylated tRNA to the D-site is dependent upon the occupancy of the P-site, suggesting that the presence of tRNA at the P-site involves changes in the properties of the ribosome.

In the same session Spirin and Serdyuk provided data indicating that the translocation step is accompanied by a relative spatial displacement of some parts of the ribosome with the amplitude of several angstroms. The translocation step which is one of the most complicated processes in the cell has been subject of wide speculations. Among different models which have been proposed to explain the molecular mechanism involved in the translocation, the one formulated by Spirin explains translocation in terms of moving apart(unlocking) of the two coupled ribosomal subunits. The model as such postulates conformational changes of considerable extent in ribosomes. However, the use of ribosome populations actively involved at a certain elongation step (i. e., either at pre- or posttranslocation stage) is a prerequisite for an unequivocal demonstration of the postulated changes. Such ribosomes have been prepared on poly(U) Sepharose columns. Electron microscopic analysis has failed to detect, however, the postulated conformational change accompanying the transition from the pre- to posttranslocation state on these ribosomal preparations. Sedimentation analysis has, on the other hand, revealed a difference in sedimentation coefficients of about 1S between the two state ribosomes, posttranslocation state ribosomes appearing less dense than ribosomes of the preceeding state. Nevertheless, the possibility has not been completely ruled out by this approach that the presence of a second tRNA on the pretranslocation state ribosome contributes to this difference. Final proof of the conformational difference is provided by neutron scattering experiments showing that the neutron radii of gyration of the ribosomal particles in the posttranslocation state are greater than those of the pretranslocation state ribosomes and that the difference observed in R_g is not due to a contribution of an additional protein or RNA in one of the states. The ribosomal parts which due to this translocation-promoted displacement are responsible for the observed difference in the compactness of the ribosomes of two functional states remain now to be established.

Yamane discussed the implications of the observed effects of *in vivo* incorporation of α-aminobutyric acid into *E. coli* proteins. Although the amount of α-aminobutyric acid incorporated in place of valine into cellular proteins does not exceed 0.015%, its presence in the cell medium causes disappearance of certain protein bands in gel electrophoresis and decrease in the activity of at least certain (if not all) enzymes. The observed physiological and biochemical effects which are multivaried attest to the difficulties in measurements of *in vivo* translational fidelity and interpretations of the corresponding data.

The regulation of the eukaryotic protein synthesis was one of the topics of priority of the symposium. The contributions from different groups on this subject focused on the regulation of the eukaryotic initiation factor 2 (eIF-2) activity in reticulocytes. Kaempfer presented evidence favoring the concept that eIF-2 can interact directly with mRNA during protein synthesis, recognizing the initiation sequence in mRNA molecules. Differential affinity of the factor for different mRNA species can provide, according to his data, a mechanism for differential gene expression at the translational level. It will be tempting to establish the relationship of this activity of eIF-2 to that of eIF-4A and/or of cap-binding proteins which according to an alternate view are primarily involved in mRNA selction. The failure to obtain with added eIF-2 more than marginal relief of competition between different mRNA species may originate from the differences of the test systems used: as argued by Kaempfer, a highly active protein synthesizing system, i. e., a micrococcal nuclease treated reticulocyte lysate supplemented with a sufficient amount of highly active and pure eIF-2 is a prerequisite for the reproducibility of their results.

The discovery of the polypeptide complex, eRF, required for the catalytic action of eIF-2, has recently helped different research groups to work out the outlines of the eIF-2 coupled mechanisms leading to the formation of the 43S preinitiation complex. Nevertheless, considerable confusion continues to exist in respect to the designation of this protein factor (a partial survey of the respective terminology for this protein is given by Matts, this volume) and time for an agreement on a uniform name seems to have arrived. Voorma discussed the mechanism of action of eRF which in the presence of GTP readily regenerates eIF-2 from the eIF-2.GDP complex. This exchange reaction is considerably stimulated by 40 S subunits. Two alternate explanations are given by Voorma for the stimulatory action of 40S subunits: fulfilling a function similar to that of prokaryotic EF-Ts, eRF may mediate the formation of eIF-2.GTP. Met-tRNA ^{fMet} complex where 40S subunits pull the reaction to the completion due to the subsequent formation of the 43S preinitiation complex; alternatively, eIF-2.eRF may be involved together with GTP and Met-tRNA^{fMet} in the formation of a quaternary complex which dissociates into eRF and the ternary complex eIF-2.GTP.Met-tRNA^{fMet} upon binding of the latter to the 40S subunit. The second view is supported by Voorma's results which seem to attest to the existence of the postulated quaternary complex.

eRF plays a central role also in the inhibition of protein synthesis by eIF-2α kinases which are activated in reticulocytes under different environmental stresses, e.g., heme deficiency, presence of oxidized glutathione or of dsRNA. eIF-2 phosphorylated on its α-subunit (eIF-2αP) undergoes a stable complex with eRF, then sequestering the limiting amounts of eRF normally present in the cell. Although only 30%-40% of eIF-2 molecules get phosphorylated under the action of eIF-2α kinases, this suffices to inhibit protein synthesis since eRF becomes unavailable for the regeneration of eIF-2. Matts reported that GDP is essential for trapping eRF in eIF-2-eRF complex and pointed out the regulative role of Mg^{2+} in the physiologic regeneration of eIF-2 from eIF.2.GDP complex.

eIF-2α kinases appear to correspond to the final stage of a cascade system, described with names like heme-controlled translational inhibitor, rather out of historical reasons. However, the initiation of activation of the system is by no means confined to heme deficiency, as has been shown by de Haro's presentation. That the eIF-2 kinases can be activated also in the presence of Ca^{2+} and phospholipid reveal another till now unknown aspect of this control mechanism which may be effective on the membrane-bound ribosomes. A detailed picture of the activation process of (and of the components involved in) the cascade system is still missing. Present state of evidence allows solely a differentiation into reversible and irreversible stages of activation. Hence, as demonstrated by Kan, the activation of an inhibitor protein of 23000 in the presence of oxidized glutathione and under high pO_2 (and possibly identical with that activated in the high pressure exposed or gel filtered lysates) may provide a clue helpful for the activation of eIF-2α kinases. On the other hand, the GDP- and ADP-dependent modulation of EF-2 activity may reflect a control mechanism coupling the elongation rate with the energy change of the cell (Bermek).

The discovery of archaebacteria has made a revision of the classical concept of evolution necessary. With the investigations of the biochemical properties of these organisms a more appropriate picture of evolution seems to be arising. Work has already revealed, as surveyed by Vazquez, some interesting features of the protein synthesizing system in these bacteria concerning sensitivity to antibiotic and toxin inhibitors. In this respect archaebacterial ribosomes have features common with both (or either) eubacterial and (or) eukaryotic ones. This may be due to that each of the three primary kingdoms of organisms, though being distinct from the other two, has origin like the others from a common ancestor. Although similar to its eubacterial counterpart in respect to its sedimentation coefficient the small subunit of the archaebacterial ribosome displays the typical eukaryotic "bill" structure adjacent to the L7/L12 stalk and to the binding site of EF-2. This might explain in the light of similarities in the amino acid sequences of L7/L12, the reported sensitivity of the archaebacterial EF-2 to diphtheria toxin.

The presentation by Şimşek covering also the topic of cell evolution provided some interesting features of the archaebacteria in the example of the gene and mRNA for a membrane protein of *Halobacterium halobium*. Particularly noteworthy is that this mRNA appears to carry only 2-3 nucleotides before the initiating AUG codon, hence lacking the prokaryotic type promotor or Shine-Dalgarno consensus sequence. Moreover, a marked preference for C or G in the third position of codons is observed during the translation of this mRNA.

Subject Index

W. Saenger

Principles of Nucleic Acid Structure

1984. 227 figures. XX, 556 pages
(Springer Advanced Texts in Chemistry)
ISBN 3-540-90762-9

Contents: Why Study Nucleotide and Nucleic Acid Structure? – Defining Terms for the Nucleic Acids. – Methods: X-Ray Crystallography, Potential Energy Calculations, and Spectroscopy. – Structures and Conformational Properties of Bases, Furanose Sugars, and Phosphate Groups. – Physical Properties of Nucleotides: Charge Densities, pK-Values, Spectra, and Tautomerism. – Forces Stabilizing Associations Between Bases: Hydrogen Bonding and Base Stacking. – Modified Nucleosides and Nucleotides; Nucleoside Di- and Triphosphates; Coenzymes and Antibiotics. – Metal Ion Binding to Nucleic Acids. – Polymorphism of DNA. – Polymorphism of DNA versus Structural Conservatism of RNA: Classification of A-, B-, and Z-Type Double Helices. – RNA Structure. – DNA Structure. – Left-Handed, Complementary Double Helices – A Heresy? The Z-DNA Family. – Synthetic, Homopolymer Nucleic Acid Structures. – Hypotheses and Speculations: Side-by-Side Model, Kinky DNA, and „Vertical" Double Helix. – tRNA - A Treasury of Stereochemical Information. – Intercalation. – Water and Nucleic Acids. – Protein-Nucleic Acid Interaction. – Higher Organization of DNA. – References. – Index.

Springer-Verlag
Berlin
Heidelberg
New York
Tokyo

R. F. Schleif, P. C. Wensink

Practical Methods in Molecular Biology

1981. 49 figures. XIII, 220 pages
ISBN 3-540-90603-7

Contents: Using *E. coli.* – Bacteriophage Lambda.
– Enzyme Assays. – Working with Proteins. –
Working with Nucleic Acids. – Constructing and
Analyzing Recombinant DNA. – Assorted Laboratory Techniques. – Appendix I: Commonly
Used Recipes. – Appendix II: Useful Numbers. –
Bibliography. – Index.

R. K. Scopes

Protein Purification

Principles and Practice

1982. 145 figures. XIII, 282 pages
(Springer Advanced Texts in Chemistry)
ISBN 3-540-90726-2

Contents: The Enzyme Purification Laboratory. –
Making an Extract. – Separation by Precipitation.
– Separation by Adsorption. – Separation in Solution. – Maintenance of Active Enzymes. – Optimization of Procedures and Following a Recipe. –
Measurement of Enzyme Activity. – Analysis for
Purity; Crystallization. – Appendix A. –
Appendix B: Solutions for Measuring Protein
Concentration. – References. – Index.

Springer-Verlag
Berlin
Heidelberg
New York
Tokyo